I0067094

William Adams

A New Operation

For Bony Anchylosis of the Hip Joint with Malposition

William Adams

A New Operation
For Bony Anchylosis of the Hip Joint with Malposition

ISBN/EAN: 9783337090104

Printed in Europe, USA, Canada, Australia, Japan

Cover: Foto ©berggeist007 / pixelio.de

More available books at **www.hansebooks.com**

A

NEW OPERATION

FOR

BONY ANCHYLOSIS OF THE HIP JOINT

WITH

MALPOSITION OF THE LIMB

BY

SUBCUTANEOUS DIVISION OF THE NECK OF THE THIGH-BONE.

BY

WILLIAM ADAMS, F.R.C.S.,

SURGEON TO THE ROYAL ORTHOPÆDIC AND GREAT NORTHERN HOSPITALS;
LATE PRESIDENT OF THE HARVEIAN SOCIETY;
LATE VICE-PRESIDENT OF THE PATHOLOGICAL AND MEDICAL SOCIETIES OF LONDON;
FORMERLY LECTURER ON SURGERY AT THE GROSVENOR PLACE SCHOOL OF MEDICINE,
AND DEMONSTRATOR OF MORBID ANATOMY AT ST. THOMAS' HOSPITAL.

Illustrated by Numerous Wood-Engravings.

LONDON:
J. AND A. CHURCHILL, NEW BURLINGTON STREET.
1871.

PREFACE.

In consequence of the great attention which the operation of "subcutaneous division of the neck of the thigh-bone" has attracted, since I first proposed and performed it on December 1st, 1869, as a means of rectifying the extreme deformity occasionally met with in cases of bony anchylosis of the hip joint, I have reprinted in a connected form the papers published in reference to this operation, the interest of which will, I trust, not be diminished by the fact that the subject matter of the present pamphlet has for the most part appeared in the journal literature of the day—see *British Medical Journal*, December 24th, 1870, and subsequent numbers.

I brought the details of the first successful case before the notice of the British Medical Association at the meeting at Newcastle in August 1870, having previously exhibited the patient at the Medical Society of London on April 25th, 1870, when he was able to stand and walk about the room without any assistance. As far as possible I have now collected the details of all the cases, published and unpublished,

in which the operation has been performed up to the
present date, so that its success or failure may be
placed fairly before the profession. So far as I am
aware, the operation has been performed six times
successfully, by five different surgeons, in cases of
bony anchylosis of the hip joint with extreme deformity, in adults; and once unsuccessfully in a case
of fibrous anchylosis with extreme deformity in a
child.

Such a result, it will be admitted, fairly establishes
the claim of the operation to be recognised as a
legitimate surgical procedure, in appropriate cases,
and will, I trust, entitle it to the further confidence of
the profession.

In the accompanying Plates I have given accurate
drawings, made by Mr. D'Alton, of all the specimens
of bony anchylosis of the hip joint (eight in number)
in the Museum of St. Thomas' Hospital, for the purpose of showing that with a variable amount of destruction of the head of the bone, in some cases, and
none whatever in others, the operation of dividing
the neck of the femur could be performed in five out
of the eight cases. For observations on this subject
see the paper on the selection of cases, &c., page 19.
I have also added woodcut (Plate IV) copied from
Cheselden's celebrated work " Osteographia," showing
bony anchylosis of the hip joint, without any destruction of the head of the bone.

These specimens were necessarily drawn in the
position in which they are put up in the bottles,
generally to show the nature and extent of the anchylosis, so that they give no idea of the amount of
deformity from the malposition of the femur in refe-

rence to the pelvis, but the direction in which the limb has become anchylosed is stated in the description of each specimen.

I have also added a short account of the operations for the division of bones and anchylosed joints previously performed in Germany, France, and America, based more or less upon the subcutaneous principle, and described collectively as subcutaneous osteotomy. For this account I am indebted to my friend Dr. Henry Dick, of whose assistance I have been glad to avail myself, from his acquaintance with the German and French literature, and also from the great interest he has always taken in subcutaneous surgery.

In conclusion, I regret being obliged to add as an Appendix to the present paper, a correspondence relative to a claim of priority in the performance of the operation of subcutaneous division of the neck of the thigh-bone which has been set up by my colleague, Mr. Brodhurst, one of the assistant-surgeons to the Royal Orthopædic Hospital, in his recently published work " On Deformities."

This correspondence has already appeared in the pages of the *British Medical Journal* on the 18th February, 1871, and in subsequent numbers, from which it is now reprinted. By reference to letters Nos. 1 and 3 it will be seen that the operation performed in 1865, upon which Mr. Brodhurst at page 152 of the work referred to, appeared first to rest his claim was performed on the great toe, and was not even of a true subcutaneous character; and that the operation performed in 1861, upon which in a subsequent letter (No. 2 of the correspondence) he seeks to claim more distinctly the subcutaneous division of the neck of the thigh-bone, was an operation partaking more of the character of excision of the hip joint, since it was performed in a case of progressive disease, dead bone being removed by the gouge, and the external incision, which at first was two and a half inches in length, being subsequently extended to five inches, (see No. 3 of the correspondence.)

The correspondence must be allowed henceforth to speak for itself, and I can only regret the necessity of reprinting it, in vindication of my own claim to the priority in the performance of this operation.

W. ADAMS.

Henrietta Street,
    Cavendish Square,
        July, 1871.

B.

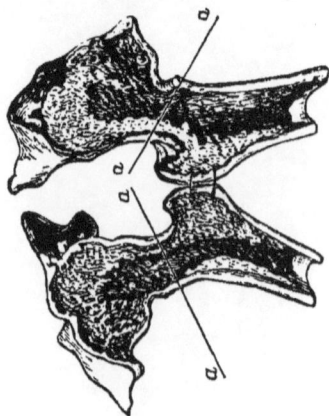

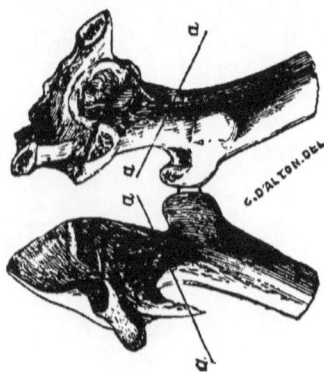

A.

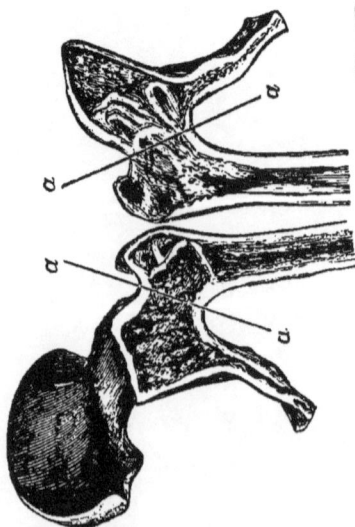

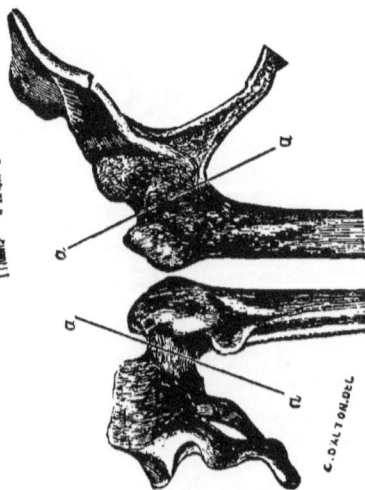

Fig. A.—The following description is from the Hospital Catalogue, "a right hip joint, showing complete bony anchylosis ; a section has been made through the bones from side to side. Externally the form of the joint is but little, if at all, altered : the margin of the acetabulum may be traced without much difficulty ; and the neck of the femur is of its natural proportions. The cut surfaces show such intimate union that the crusts and cancellous tissues of the bones are continuous ; and it is impossible to distinguish their boundaries. The bones are very heavy ; and their crust is very compact and ivory-like."* Very little new bone has been thrown out externally in the neighbourhood of the anchylosis.

In this specimen the shaft of the femur is flexed nearly at a right angle directly forwards, with slight abduction.

The operation of dividing the neck of the thigh-bone could easily have been performed in the situation indicated by the line *a a.*

Fig. B.—Bony anchylosis of left hip joint. A section of the specimen has been made and is shown in the Plate. The cancellous structures of the two bones are continuous, so that the outline of the joint cannot be traced on the section.

There appears to have been no destruction whatever of the head of the bone ; and the neck of the femur is of its full natural proportions. Indeed the neck of the femur in this specimen appears to be of abnormal length, in consequence of the abducted position of the limb. The pelvic bones, however, have been so much cut away in preserving this specimen, that there might be some difficulty in determining the precise direction in which the limb was fixed, though it appears to have been flexed at about an angle of 45 degrees, and abducted. Some new bone has been thrown out externally from the surface of the ilium, in the neighbourhood of the anchylosis, anteriorly.

The operation of dividing the neck of the thigh-bone could easily have been performed in the situation indicated by the line *a a.*

This is one of the old specimens in St. Thomas' Museum ; but does not appear to have been described in the new catalogue.

---

* St. Thomas's Hospital Museum Catalogue, Vol. II., Pathological Anatomy, No. 51D.

A.

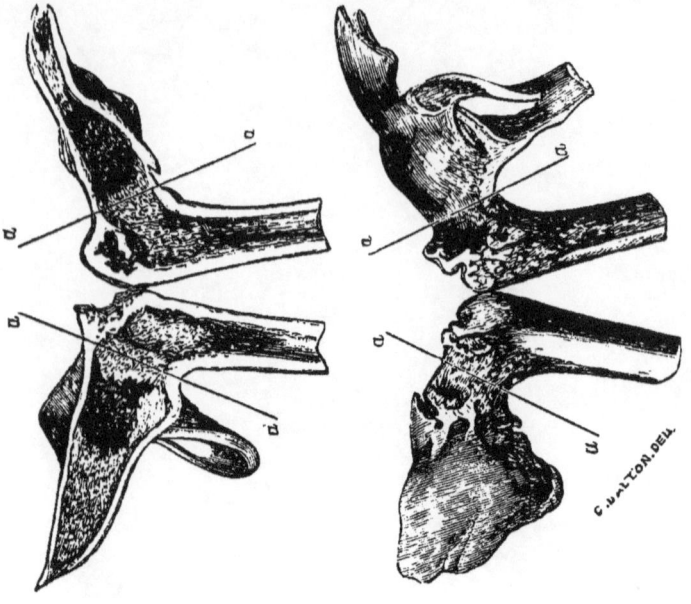

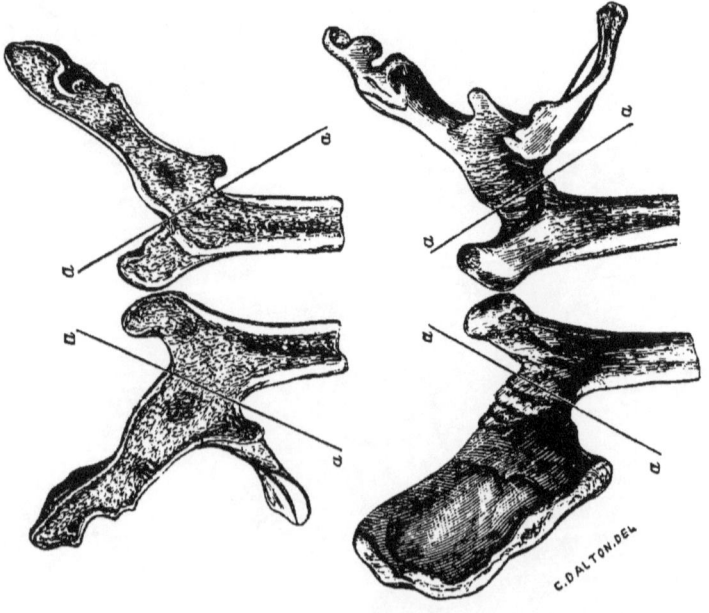

C.DALTON.DEL.

C.DALTON.DEL.

DESCRIPTION OF PLATE II.

Fig. A.—Bony anchylosis of left hip joint. A section of the specimen has been made, and is shown in the Plate. "The cancellous structures of the two bones are continuous ; so that no trace of the situation of the joint can be found in the section."* The head of the bone has been destroyed to a considerable extent; but the neck still remains, though apparently shortened from loss of substance, in the head of the bone. Very little new bone has been thrown out externally in the neighbourhood of the anchylosis.

In this specimen the shaft of the femur is flexed at a right angle directly forwards, without abduction or adduction.

The operation of dividing the neck of the thigh-bone could easily have been performed in the situation indicated by the line *a a*.

Fig. B.—Bony anchylosis of left hip joint, a section of the specimen has been made, and is shown in the Plate. "The neck of the femur is somewhat shortened, its trochanters but little developed, (destroyed? W. A.), and the trochanteric pit very shallow. The cut surfaces show no trace of the head of the femur; the crust of its neck is perfectly continuous with that of the os innominatum."† New bone has been thrown out externally in the neighbourhood of the anchylosis, and, as stated in the catalogue, "periosteal deposit has occurred in several parts of the ilium."

In this specimen the shaft of the femur is flexed at a right angle, and slightly adducted.

The operation of dividing the neck of the thigh bone could easily have been performed in the situation indicated by the line *a a*.

---

* St. Thomas' Hospital Museum Catalogue Vol. II, Pathological Anatomy, No. 53D, presented to the Museum by Carr Jackson, Esq.

† St. Thomas' Hospital Museum Catalogue, Vol. II, Pathological Anatomy, No. 52D.

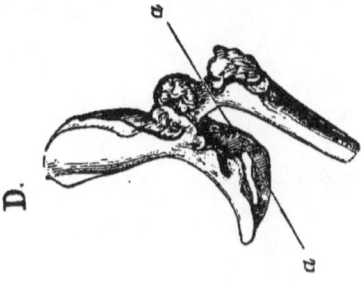

D.

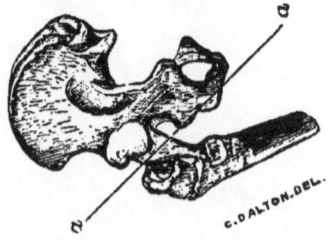

E.

C.DALTON.DEL.

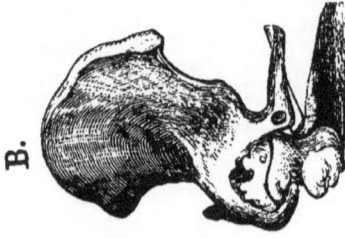

B.

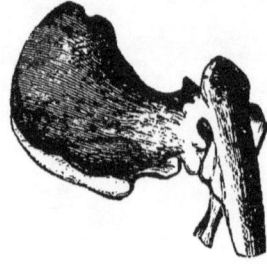

C.

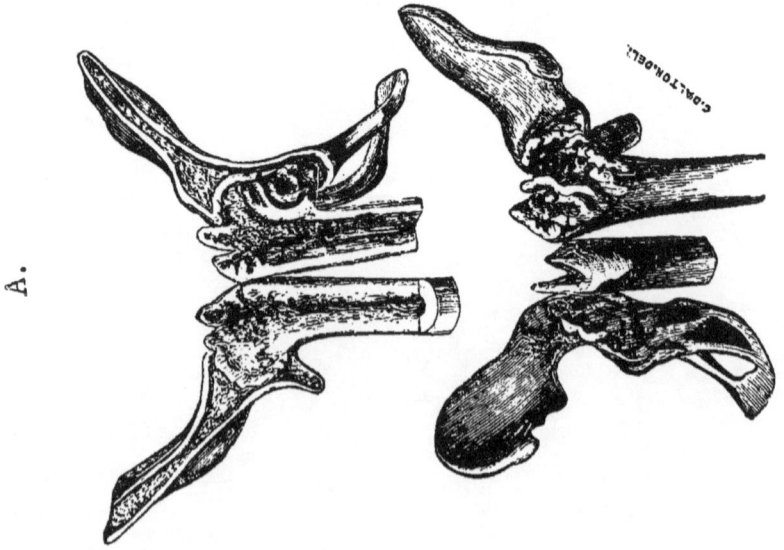

A.

C.DALTON.DEL.

Fig. A.—Bony anchylosis of right hip joint. A section of the specimen has been made, and is shown in the Plate. The following description is from the Hospital Catalogue : " A right os innominatum and the upper part of the corresponding femur, showing hip disease progressing towards anchylosis. The section that has been made shows considerable destruction of the head of the femur ; its upper surface is connected with the acetabulum by a quantity of loose spongy bone ; and below, a dense bony bridge connects the lower margin of the acetabulum with the upper part of the shaft of the femur. Between these two points, however, a considerable interval exists between the femur and the acetabulum (best seen in the anterior section) ; and caries seems still to have been in progress on the surface of the former. Behind the seat of anchylosis great destruction of the ischium has also occurred ; in one place extending quite through the bone to the cavity of the pelvis. In some parts this destroyed surface has undergone repair ; in others it seems still carious. The cancellated tissue of the great trochanter has been removed ; whether by caries, necrosis, or accident, is uncertain. The femur is much increased in thickness and density."*

In this specimen the shaft of the femur is flexed directly forwards at an angle of 45⁰, without any appreciable abduction or adduction.

The operation of dividing the neck of the thigh-bone could not have been performed in consequence of the extensive destruction of the head and neck of the bone.

Fig. B.—Bony anchylosis of right hip joint incomplete, with necrosis still existing in the head of the bone, and also in the acetabulum, communicating with the pelvic cavity ; *a.* line of section made through the head of the bone in removing the specimen. A small circumscribed necrosis also exists in the pubic bone.

In this specimen the shaft of the femur is flexed at a right angle, and adducted to an extreme degree, so as to narrow the angle between the thigh-bone and the pubis to a degree which would render the operation of dividing the neck of the thigh-bone impracticable.

A recent specimen not described in the published catalogue. No. 53¹ in MS. catalogue.

Fig. C.—Bony anchylosis of left hip joint. The following description is from the Hospital Catalogue: " All traces of the acetabulum and head of the femur are lost ; and the osseous structure of the innominate is perfectly continuous with that of the femur. The anchylosis is evidently of old date ; the surface of the uniting bone is smooth and dense, and apparently everywhere healthy. A considerable quantity of new bone has been thrown out at the seat of anchylosis.

In this specimen the shaft of the femur is flexed nearly at a right angle, and adducted.

The operation could not have been performed in consequence of the destruction of the head and neck of the bone, and fusion of the femur with the pelvic bones in the situation of the joint.†

Figs. D and E.—Anterior and posterior views of bony anchylosis of left hip joint incomplete, with progressing disease. and partial displacement upwards of head of bone in a young subject. New bone has been thrown out in the line of the acetabulum at the upper part, where anchylosis is complete.

In this specimen the shaft of the femur is slightly flexed, but adducted to a greater extent.

The operation of dividing the neck of the thigh-bone could easily have been performed, though as disease was still progressing at some parts, it would not have been indicated.

Recent specimen not described in published catalogue. No. 53² in MS. catalogue.

---

* " St. Thomas' Hospital Museum Catalogue." Vol. II. Pathological Anatomy, No. 48D.

† " St. Thomas' Hospital Museum Catalogue." Vol. II. Pathological Anatomy, No. 51'D.

PLATE IV.

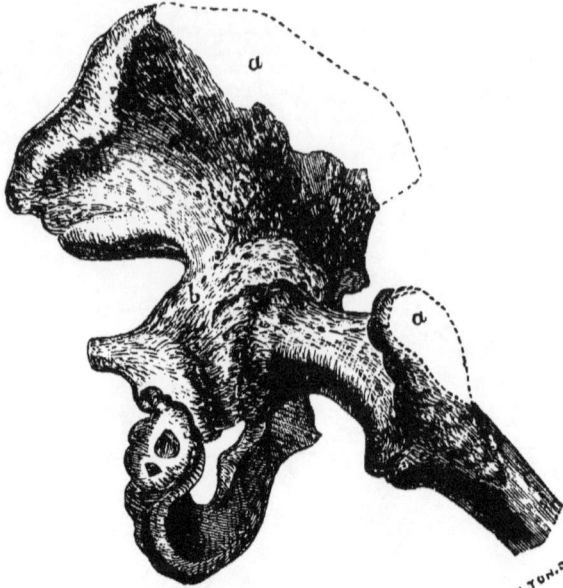

C. D'AI.TON.DEL.

DESCRIPTION OF PLATE IV.

Bony anchylosis of right hip joint, with little, if any, destruction of the head of
the bone, and the neck of the femur of its natural proportions, so that the
operation of dividing the neck of the bone could easily have been performed.
The shaft of the bone considerably abducted, and in a flexed position.
Copied from an engraving, Tab. 47, Fig. 1, in the "Osteographia," by William
Cheselden, F.R.S., London, 1733   In Chapter VII, after a few observations on
diseases of the joints, Mr. Cheselden observes: "Sometimes in these cases the
ends of the bones erode, then joyn together and form an anchylosis (Tab. 47),
which though a bad disease of itself, yet it is often a remedy of this disease,
which is much worse."

In the description of Plate 47, Mr. Cheselden observes: "An anchylosis of
the os innominatum and os femoris, communicated to me by Mr. Westbrook.
a a. Part that was broken off.   *b*. A part rough and carious."

(library stamp) BOSTON
JAN

REMARKS

ON THE

SUBCUTANEOUS DIVISION OF THE NECK OF THE THIGH-BONE,

AS COMPARED WITH OTHER OPERATIONS FOR RECTIFYING
EXTREME DISTORTION AT THE HIP JOINT, WITH BONY
ANCHYLOSIS.*

CASES of extreme deformity at the hip joint, with
the thigh flexed upon the pelvis, and generally adducted or drawn towards the opposite limb, sometimes
even crossing over the opposite limb, are not un-
commonly met with in surgical practice as the result
of various forms of hip joint inflammation.

Anchylosis of the hip joint may be either true or
false as in other articulations. By true anchylosis, I
mean bony union of the articular surfaces after complete destruction of the joint; and by false anchylosis,
either one of two conditions; viz., 1. Union of the
articular surfaces by fibrous tissue after ulceration of

---

* Read before the Surgical Section at the Annual Meeting of the
British Medical Association in Newcastle-upon-Tyne, August 1870, and
reprinted from the *British Medical Journal*, December 24th, 1870.

B

the articular cartilage, and partial or complete destruction of the joint; or, 2. Inflammatory thickening
and retraction of the ligamentous and other fibrous
structures external to the joint; the joint itself remaining in a healthy, or nearly healthy condition,
without any destruction of the articular cartilages,
but sometimes with intracapsular adhesions.

In the present paper it is not my intention to speak
of *false anchylosis;* but, in reference to the first condition of false anchylosis above described, I would
only observe that it is generally the result of strumous
disease, with ulceration of the articular cartilage, and
that any amount of contraction and deformity may be
overcome by gradual mechanical extension, with or
without tenotomy, according to circumstances, so as
to bring the limb into an improved position ; but there
never can be any reasonable expectation of restoring
motion in the joint, and forcible extension is extremely hazardous, as liable to set up inflammation and
re-excite destructive disease in the articular extremi-
ties of the bones.

With regard to the second condition of false anchylosis, in which there is no destruction of the articular
cartilages, I would only observe that this condition
is generally the result of acute rheumatism, frequently
with gonorrhœal complication, or, as it is called,
gonorrhœal rheumatism. In this form, forcible extension under chloroform is especially applicable, any
contracted tendons having been divided three or four
days previously, where this may appear necessary ;
by it the deformity may be completely overcome and
free motion of the joint restored in a large proportion
of cases, if the treatment be commenced within six

months or a year of the rheumatic inflammation. Occasionally I have myself succeeded in restoring motion
at a much later period, even as late as three years;
but, on the other hand, I have entirely failed after
the lapse of a year, such cases passing into complete
bony anchylosis. Gradual mechanical extension, combined with passive motion, will occasionally succeed
in this class of cases when commenced early, but is
very liable to failure.

Either of these two forms of false anchylosis may
result from traumatic inflammation occurring in a
healthy individual; but unless suppuration in the joint
occurs, the latter form generally takes place, the articular cartilages remaining in a healthy condition; and
in such cases forcible extension under chloroform,
with or without tenotomy, according to circumstances,
generally gives the most satisfactory results.

True bony anchylosis can only occur after complete
destruction of the joint and removal of the articular
cartilages. This condition may be the result either
of the strumous, the traumatic, the more severe
forms of rheumatic, or the pyæmic form of inflammation. When resulting from strumous disease after
. ulceration of the articular cartilages and caries of
bone, it occurs only at a late period—generally from
seven to ten years after disease has ceased. I have
known fibrous anchylosis to remain in the hip joint
seven years after active disease had ceased; and it is
an admitted fact that bony anchylosis is very slow to
result in this class of cases. When resulting from
traumatic inflammation, bony anchylosis may take
place within a year if suppurative inflammation of
the joint has occurred; but, if not, the articular car-

tilages remaining sound long after the injury, bony
anchylosis is not to be looked for, except perhaps as
a late result after many years. When resulting from
the more severe forms of rheumatism—especially the
gonorrhœal form—bony anchylosis generally occurs in
a period of from one to three years; the articular
cartilage being first covered by the lymph effused
during the inflammatory process, and then slowly
disappearing after membranous bands of adhesion
have been formed between the adjacent articular
surfaces; i.e., from one articular cartilage-surface to
the other. For an illustration of this form, see case
recorded by myself in the Pathological Society's
Transactions, vol. xx, p. 296; London, 1869. When
resulting from acute pyæmic inflammation, causing
rapid and complete destruction of the joint, bony
anchylosis occurs more quickly than after any other
form of inflammation, except when acute suppuration
follows as the result of an injury in a healthy individual. In the pyæmic as well as in the traumatic
form, bony anchylosis may occur within the period of
a year.

When bony anchylosis of the hip joint has taken
place as the result of any of the inflammatory affections above described, no operative procedures should
be attempted, if the anchylosis have occurred with
the limb in a straight position, as any attempt to
obtain free motion by the production of an artificial
joint could only be made at the risk of life, and under
the most favourable circumstances with a very doubtful result as to useful motion. In the great majority
of cases of bony anchylosis of the hip joint, however,
contraction of the joint with the limb in a deformed

position is found to exist; in some cases, simple flexion of the thigh having occurred with very little adduction or abduction; in other cases, severe adduction, with a comparatively small amount of flexion; and again, in others, the distortion will be found to depend upon flexion with abduction, or adduction, of the thigh, in about equal proportions.

The inconveniences consequent upon anchylosis of the hip joint with distortion will vary according to the extent, and also according to the direction, in which the distortion has occurred; e.g., in a case of simple flexion of the thigh, even when contracted to a right angle with the pelvis, and accompanied with abduction, as in the case in which I subcutaneously divided the neck of the thigh-bone, and the photographs of which I now exhibit to the Meeting, the inconvenience was limited to the uselessness of the limb, the use of a crutch and stick being necessary for progression.

In cases of anchylosis with distortion, in which adduction of the thigh predominates, so that the knee is drawn across the opposite thigh; or in cases in which the adduction and flexion are combined, the inconveniences are very great when occurring in females, in consequence of pressure upon the labia and the orifice of the vagina rendering urination difficult; and the thigh liable to constant excoriation. This occurred to a serious extent in one of the cases operated upon by Louis Sayre of New York; and in this case even the introduction of a catheter was found impossible. Inconvenience of a similar kind occurred in the case of a young lady who was the subject of fibrous an-

chylosis of the hip joint, with the thigh in an extremely adducted position, with very little flexion,
and upon whom I have operated with partial success
by means of forcible extension, under chloroform,
after tenotomy.

These and other inconveniences will be found to
depend upon the extent and direction in which the
distortion has occurred, and it is therefore obvious
that in some cases an urgent necessity exists for
surgical interference ; and various operations have
been proposed not only with the object of rectifying
the deformity and bringing the leg into a straight
and useful position, but it has also been attempted
to obtain free motion by the production of an artificial joint.

The first operation, having for its object not only
that of rectifying the deformity, but of obtaining
motion by the establishment of a false joint, was
performed by Dr. Rhea Barton* of Philadelphia,
United States, in 1826. This operation was accomplished by a crucial incision made over the great trochanter, seven inches in length and five inches in a horizontal direction. The bone was then divided transversely by a fine saw—it is said "between the two
trochanters"—probably just above the small trochanter. The natural direction of the limb was at
once restored, and the case proceeded favourably.
It is said that useful motion was obtained, but that
seven years afterwards anchylosis took place, and that

---

* " On the Treatment of Anchylosis by the Formation of Artificial
Joints," in the *North American and Surgical Journal*, April 1827; with
further remarks in the *American Journal of the Medical Sciences*, vol.
xxi.

the man died of phthisis nine years after the operation.

The next operation worthy of attention is recorded by Dr. Louis Sayre* of New York, who operated successfully on two cases in which he performed a new operation, which he had proposed for obtaining a false joint and preserving motion in cases of bony anchylosis of the hip joint, with the thigh in a flexed position. The theory of this operation was to obtain free motion by the formation of a false joint, of a ball-and-socket character, supposed to resemble the hip joint in possessing an acetabulum or cavity corresponding to this, and a rounded extremity of bone corresponding to the head of the femur; and also a round ligament.

FIG. 1.   FIG. 2.

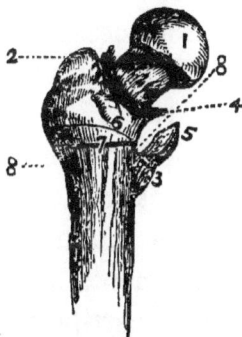   

Fig. 1.—Diagram copied from Louis Sayre's pamphlet. 1. Head of Femur. 2. Trochanter Major. 3. Trochanter Minor. 4. Line of insertion of Capsular Ligament (variable.) 5. Tendon of Psoas Magnus and Iliacus Internus Muscle. 6. Line of Curved Section. 7. Line of Transverse Section. 8, 8. Dotted Lines indicating rounding off of Lower Fragment after removal of Segment.

Fig. 2.—Upper portion of Thigh-bone. Situation and direction of Subcutaneous Division of Neck of Thigh-bone proposed by Mr. Adams, represented by line *a, a*.

* "A New Operation for Artificial Hip-Joint in Bony Anchylosis ; illustrated by Two Cases." By Lewis A. Sayre, M.D. New York: D. Appleton and Co. 1869.

The operation consisted in the removal of a transverse section of the femur, of elliptical form, just above the trochanter minor (as shown in Fig. 1), by means of the chain-saw, an incision of about six inches in length being made over the trochanter major in the axis of the limb. The first patient, Robert Anderson, aged 26, was operated upon on the 11th June, 1862, and in December of the same year he is reported as follows, " Could stand on either leg without either crutch or cane;" and as late as April 29th, 1868, Dr. J. S. Green says, in a letter to Dr. Sayre, that "Robert Anderson still lives, moves, and walks with practical agility," (p. 35, pamphlet).

The second operation was performed 6th November, 1862, on Miss Susan M. Losee, aged 24. This case proceeded less favourably than the first; but all discharge from the wound ceased four months after the operation. Subsequently, however, an abscess formed, and a little necrosed bone escaped. Pneumonia and pleurisy occurred, and she died on the 17th May, 1863. At the *post mortem* examination, tubercular deposits were found in the lungs, and a large abscess in the left lung. The artificial joint was found to be provided with a complete capsular ligament, and the articulating surfaces were tipped with cartilage and furnished with synovial membrane. In consequence of Dr. Bauer, in his work on *Orthopœdic Surgery* (p. 325), stating that this case of Dr. Sayre's died of pyæmia, a number of letters from medical men are given in the appendix to the paper, confirming the tubercular theory, and also Dr. Sayre's statement as to the existence of cartilaginous covering to the bone, synovial membrane, &c.

I am not aware of any operation having been performed in this or any other country on the hip joint in cases of bony anchylosis, with the object of obtaining free motion by the establishment of a so-called " artificial joint ;" and it will at once be seen that further experience is required before we can confidently speak of the success of such an operation, although the cases recorded by Dr. Sayre are undoubtedly worthy of the most attentive consideration.

It occurred to me, however, that in these cases of bony anchylosis of the hip joint, with extreme distortion, a much more simple operation might be performed by the subcutaneous division of the neck of the thigh-bone, about its centre, within the capsular ligament (as represented in Fig. 2) ; and on the 1st December, 1869, I performed this operation successfully on the following case.

Luke Bristowe, aged 24, a gardener from Loudon, near Chippenham, Wiltshire, was admitted into the Orthopædic Hospital on the 12th October, 1869, in consequence of extreme deformity at the hip-joint. The thigh was flexed upon the pelvis at a right angle, and firmly anchylosed in this position ; the heel of the right leg rested on the upper part of the left knee joint, and the limb was therefore perfectly useless. The only mode of progression was either with two crutches, or with one crutch and a stick, which he generally used (as shown in Fig. 3). He had also anchylosis of the vertebral articulations through a considerable portion of the spinal column ; all the

lumbar and lower dorsal vertebræ were perfectly im-
moveable, and the spine was curved posteriorly, with

FIG. 3.

Fig. 3.—Position of Limb previous to operation, and the usual mode of progression.

an inclination to the right side in the lower dorsal
and upper lumbar region. In consequence of this
anchylosis through the lumbar region, the pelvis and
spine could only be moved together, and the trunk
and leg therefore appeared to be remarkably fixed in
the deformed position. There was also a fixed and
permanent obliquity of the pelvis, with regard to
the spinal column, to the extent of two inches, as
ascertained by careful measurement. Partial anchylosis also existed in some of the upper cervical vertebræ; the motion between the occipital bone and the

atlas was free, but between the atlas and the axis motion was extremly limited ; and the head was habitually carried forwards. This affection was the result of an extremely severe attack of rheumatic fever, with which he was seized seven years ago June last, and which, he stated, was not preceded by gonorrhœa. Various articulations were affected during the fever, and the rheumatic pains were severe for six months, and continued more or less for twelve months. During the latter part of this time he was an inmate of the Bath Hospital for fifteen weeks, and had the hot mineral baths, but without any marked relief; and he then went into the Brighton Hospital for nine weeks, where he was galvanized, and had to swing a seven-pound weight, but without material benefit. No treatment had been adopted during the last six years, nor had he suffered from any further attack of rheumatism. His general health was good, and also his family history.

That the case was one of true bony anchylosis, was proved by the failure of forcible extension under chloroform, tried on three separate occasions at the Orthopædic Hospital; and after this I suggested to the patient the operation of dividing through the bone as the only means of restoring the position of the limb, and he readily gave his assent. For the purpose of the operation, and that he might have the advantage of superior nursing, if required, he was removed to the Great Northern Hospital on the 26th November, 1869.

On the 1st December, 1869, I divided the neck of the thigh-bone subcutaneously within the capsular ligament, in the presence of my colleagues, Mr.

.Gay, Mr. Carr Jackson, and Mr. Shillitoe; Mr. Mason, Dr. H. Dick, and Mr. John Mackenzie of the Bombay Hospital, were also present. The instruments used were a long tenotomy-knife, and a very small saw, three-eighths of an inch in width, and with a cutting edge an inch and a half in length, at the end of a slender shank three inches in length, made by Mr. Blaise of St. James's Street, and shown on a diminished scale in Fig. 5. The details of the operation were as follows.

I entered the tenotomy-knife a little above the top of the great trochanter, and, carrying it straight down to the neck of the thigh-bone, divided the muscles and opened the capsular ligament freely.

FIG. 4.

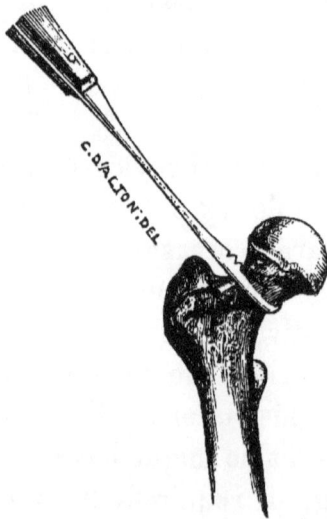

Withdrawing the knife, I carried the small saw along the track made—preserving this by pressure of the fingers—straight down to the bone, and sawed through it from before backwards, in the direction represented in Fig. 4, which shows the saw applied to the anterior surface of the neck of the bone. The section of the bone was accomplished in five minutes. No hæmorrhage followed; and I immediately applied a compress of dry lint, retained in position by strips of plaster, and a bandage.

Fig 4. Upper portion of thigh-bone showing the saw applied to the anterior surface of the neck of the bone previous to its division, from before backwards, in the situation and direction indicated.

As soon as the bone was cut through, the leg

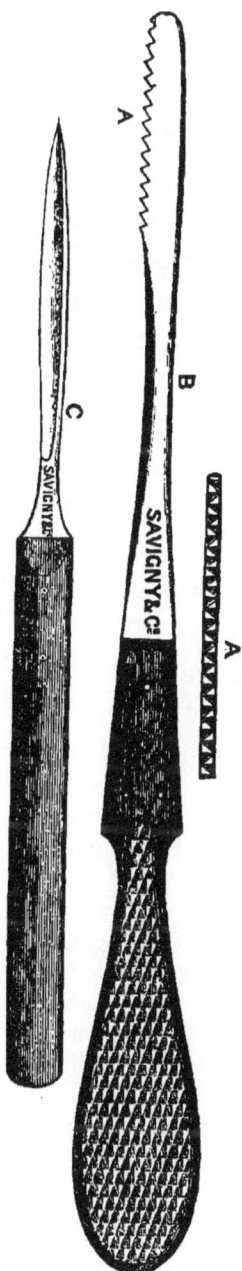

FIG. 5. FIG. 6.

Fig. 5. B, Subcutaneous Saw and C, Tenotomy-knife, drawn one-third less than those used in the operation; A representing double cutting edge of Saw, full size.

Fig. 6. Represents the Subcutaneous Saw with an improved handle, which was first made by Mr. Blaise for Mr. Jowers, of Brighton, who found that a stronger purchase in the handle than that which I had previously used (see Fig. 5) was desirable. And as Mr. Jessop, of Leeds, had also complained of the small size of the handle, I have since adopted this enlarged handle. Drawn one-third less than the full size used.

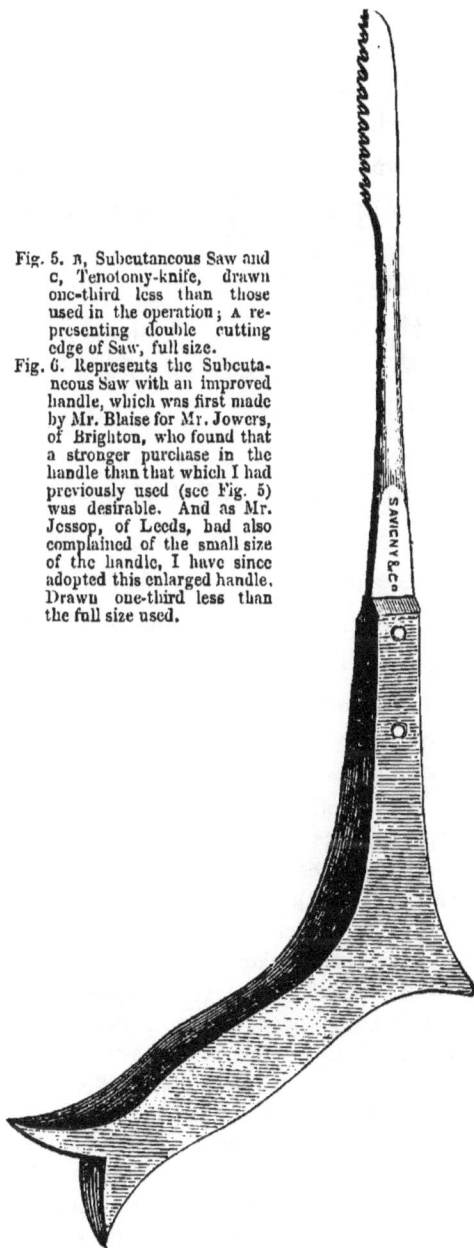

moved freely in all directions; but, before it could
be brought into a straight position, it was necessary
to divide the tendons of the long head of the rectus
and of the adductor longus muscles, and to cut
through the tensor vaginæ femoris muscle. The
limb was fixed in a straight position and bandaged
to a long interrupted Liston's splint. No inflammation whatever followed the operation; no swelling
or redness of the skin, or any deep suppuration;
but the wound healed slowly.* The House-Surgeon,
Mr. Willis, reports as follows.

December 4th. The long splint was changed to-
day for a short one. He could move the leg from
the hip gently whilst the splint was off. The tenotomy wounds were quite healed.

December 5th. A four-pound weight was attached
to the leg, which he bore well for a time.

December 7th. The dressing was removed for the
first time to-day. A few drops of pus only escaped
from the superficial wound. There was no deep

* The operation is one which, although it appears to be extremely
simple from its bloodless character and a puncture only being made,
requires to be carefully studied, especially with reference to the altered
relation of the thigh-bone to the pelvis, and the effect this may have in
some cases in narrowing the angle between the shaft of the bone, and
the pelvis; particularly in those cases in which the thigh is adducted
as well as flexed. This is the most frequent position after hip joint disease,
when in consequence of partial destruction of the head of the bone this
angle is often so contracted as to render the operation very difficult,
and even in some cases impossible, as shown in Plate III A. The
altered direction of the neck of the bone must also be carefully calculated by the surgeon, at the time of making the section, which
should be at right angles to the long axis of the neck, care being taken
to avoid cutting obliquely through the neck, or in a direction parallel
with the shaft of the bone—a mistake easily committed if the altered
direction of the shaft of the bone be not carefully calculated.

suppuration going on. The superficial wound was
dressed with carbolic lotion, one part in forty.

December 13th. The splint was removed altogether to-day. He could draw the leg up almost as
well as the sound one, and had very fair motion at
the hip.

December 22nd. He was going on well. No febrile symptoms or deep suppuration were going on.
The superficial wound was nearly healed. There
never had been more than two or three drops of pus
on the lint in the morning. Collodion and castor-oil
were applied to-day instead of carbolic acid lotion.
He got up to-day for the first time, just three weeks
since the operation.

I encouraged motion from the 13th December, and
moved the limb frequently myself; and, when he
walked about the ward on crutches, induced the
patient to swing the leg as much as possible. After
walking about daily for a fortnight, however, the
limb began to stiffen at the hip, and all attempts at
movement were painful. I then determined to aban-
don all idea of obtaining motion, and endeavour to
procure bony anchylosis with the limb in a straight
position.

On the 6th January, 1870, the man was ordered
to keep his bed, and the limb to be maintained in a
straight position by an extending weight of from
three to five pounds suspended over the end of the
bed. At the end of three weeks, the divided neck of
the femur seemed to be firmly anchylosed; and on
the 24th January, he was discharged and transferred
to the Royal Orthopædic Hospital.

March 18th. He had gained sufficient strength to

be able to walk about the ward with the aid of one stick, and he could walk a little without any assistance. [Fig. 7, taken from a photograph, represents him in the standing position, and a small depressed cicatrix is shown at the seat of operation.]

April 25th. He was taken to the Medical Society of London, and exhibited his power of walking about the room without any assistance. He continued steadily to improve, and gained sufficient strength to bear the entire weight of his body on the leg which had been operated upon, as shown in Fig. 8, in which he is represented as standing on this leg. He still,

FIG. 7.

FIG. 8.

Fig. 7.—Patient in standing position, between three and four months after operation. a. Cicatrix of subcutaneous puncture.
Fig. 8.—Patient standing, and bearing the weight of his body on the leg which had been the subject of operation.—Taken at the same time as Fig. 7. Drawn on wood from photographs by Mr. Mayall, who, by his artistic assistance, generously aids the Orthopædic Hospital.

however, generally uses one stick in walking, and walks with the body somewhat inclined to the right side (as represented in Fig. 7) in consequence of the obliquity of the pelvis and anchylosis of the lumbar and lower dorsal vertebræ. Although there appears to be a little shortening of the right leg, he is not at all improved, as to the erect position of his body, by any addition to the boot, so that he wears boots of the same thickness.

November 10th. This patient having been at his home in the country for several months, came to London, and I again examined him. He could now walk three or four miles with ease, and did not require a stick for walking purposes, although he always used one to improve the general balance of his body, which was somewhat disturbed by the spinal curvature.[*]

In comparing the relative merits of the three operations which I have now described for rectifying extreme distortion at the hip joint with bony anchylosis, the different objects sought to be accomplished and the risk to life incurred in each operation must be borne in mind. The operations performed by Barton and Sayre were undoubtedly of a formidable character, requiring large external incisions, and necessitating considerable disturbance of structures at a great depth from the surface, to allow of the use either of an ordinary saw, or of a chain-saw, as employed by Sayre ; and the evidence is not yet sufficient to prove that even by such means a permanently

---

[*] This patient in a letter dated May 18th, 1871, says, "I am pleased to tell you I am much stronger than when you last saw me. I can now walk seven or eight miles with ease, with the aid of one stick."

useful artificial joint can be established in the
neighbourhood of the hip joint. In Rhea Barton's
case, bony anchylosis was proved to have taken place
by the *post mortem* examination of the patient, who
died from phthisis eight years after the operation,
although motion is said to have been preserved for six
years.

In Sayre's first case, in which a segment of bone
was removed above the small trochanter, a useful
limb was permanently obtained, and good motion
existed five months after the operation, but some
necrosis subsequently occurred; and the late accounts,
six years afterwards, are less satisfactory than we
could desire as to the evidence of free motion at the
joint. In Sayre's second case, which undoubtedly
offers a good illustration of the establishment of a
false joint, necrosis, although in a limited degree, was
still proceeding at the time of death—six months after
the operation—and bone had exfoliated previously, so
that Dr. Bauer was induced to believe the death arose
from pyæmia rather than phthisis.

In the case which I have now brought before the
meeting, I was encouraged, by the absence of inflammation, to hope for the establishment of motion;
but, this failing, the result was limited to remedying
the deformity and obtaining a useful limb for the
patient, with bony anchylosis in a normal position;
and to such a result I would advise, in all future
operations, that our expectations should be limited.

With regard to the subcutaneous operation, which
so far as I know, was first suggested and performed
by myself, and which I have now brought before the
British Medical Association, I would only observe

that the subcutaneous division of bone—subcutaneous osteotomy, as it may be called—has proved itself to be as simple and harmless an operation in its immediate effects as subcutaneous tenotomy, with which, in its essential characters, the operation may be compared.

ON THE SELECTION OF CASES FOR THE OPERATION OF SUBCUTANEOUS DIVISION OF THE NECK OF THE THIGH-BONE.*

Since I brought under the notice of this Society, on April 25th, 1870, the operation of the subcutaneous division of the neck of the thigh-bone—an operation which I had performed for the first time in the annals of surgery on December 1st, 1869, and it may be remembered that I exhibited the man in this room, able to stand and walk without any assistance—the operation has excited a great deal of attention, and been successfully performed four times by different provincial hospital surgeons—viz., twice by Mr. T. R. Jessop of Leeds, once by Mr. Furneaux Jordan of Birmingham, and once by Mr. F. W. Jowers of Brighton. The operation has been only once performed in London, by Mr. J. Croft of St. Thomas' Hospital, on March 4th, 1871, on a boy aged eight years, who had fibrous anchylosis of the left hip joint, with the limb in a deformed position, the femur being flexed at a right angle and considerably ad-

* Read at the Medical Society of London, April 24th, 1871, and reprinted from the *British Medical Journal*, May 20th, 1871.

ducted. There had been abscesses in the neighbourhood of the joint, but for five years all suppuration
was said to have ceased, and the hip joint disease
was said to have commenced when the child was
about two years old. I was present at this operation,
and concurred in the advisability of its performance,
in consequence of the extreme deformity, which ren-
dered the success of any other method of treatment
very doubtful; as well as of the length of time which
had elapsed since the closing of the abscesses, although
the case belonged to a class altogether different from
that for which I had originally proposed the operation. In this case, extensive suppuration followed
the operation, and the child died of pyæmia on March
30th. The *post mortem* examination showed that the
neck of the bone had been divided. The head of the
bone had been, to a great extent, destroyed by caries,
and the fibrous anchylosis broken up by the suppurative process, and perhaps partly by some attempts
at forcible extension, which had been employed a
week previously, when the contracted muscles in the
neighbourhood of the joint had been divided.

The general result, therefore, is, up to the present
time, that this operation has been performed in six
cases : in five cases for bony anchylosis, successfully ;
and once for fibrous anchylosis in a child, unsuccessfully; and in the first class of cases, either no
suppuration, or very little suppuration followed the
operation ; whilst, in the latter, acute suppuration set
in and terminated fatally.

Amongst the various questions which have arisen in
connexion with this operation, two of the most important refer to the particular class of cases to which the

operation is applicable, and to those in which the operation cannot be performed ; and the object of the present paper will be to define these cases, for practical purposes, with as much accuracy as possible.

Bony anchylosis of the hip joint is known to take place as the result of several morbid conditions and diseases, which, in their general pathology and progress, especially in reference to their destructive or non-destructive character, as affecting the bones, are essentially different. For example, when bony anchylosis has taken place as the result of strumous disease of long standing, and accompanied with suppuration, destruction of the head of the bone from caries and necrosis, and sometimes also of the neck, to a greater or less extent, generally occurs, the disease itself being essentially of a destructive character, tending to caries and necrosis of bone. In these cases, displacement and fusion of parts in abnormal positions frequently take place.

On the other hand, when bony anchylosis has been the result of acute rheumatic inflammation, in its more severe form, as we not infrequently see it in the so-called gonorrhœal rheumatism, when it is generally localised in one joint, no disposition whatever exists either to the destruction of bone, or to dislocation : acute rheumatic inflammation not being of a destructive character, but tending to the effusion of plastic lymph, and organisation of adhesions, without any disposition to suppuration, to ulceration, or to caries in the bones. Rheumatic inflammation will, therefore, produce adhesions within the joint, and a kind of fibrous anchylosis, which, after a slow disintegration of the articular cartilage, terminates in true

bony anchylosis, but without any loss of bone structure; so that, in such cases, the head and neck of
the thigh-bone invariably remain of their full natural
proportions.

It is equally true that the head and neck of the
thigh-bone remain unaltered in size in cases of bony
anchylosis after pyæmic inflammation, especially in
its subacute form, from which patients generally
recover, and to which gonorrhœal rheumatism is by
some thought to belong. In this class of cases, the
soft structures of the joint, including the cartilage,
are destroyed, and bony anchylosis results; but there
is no fear of progressive caries, or necrosis of bone,
so that anchylosis is produced without loss of bone
substance. Examples of this class are not unfre-
quently met with as the result of puerperal phlebitis
and pyæmia occurring in the course of the affection.

In some other classes of cases the soft structures
of the joint are destroyed by acute inflammation, and
bony anchylosis is produced without loss of bone
substance. This occurs, for example, after traumatic
inflammation in healthy adults, such as that which
follows wounds of the joints, and gun-shot wounds
in the neighbourhood of the joints, when the joint
itself has escaped injury, and, in some cases of anchylosis, chiefly from long-retained position.

This is an important pathological law; and, as the
case upon which I operated was one of bony anchylosis after acute rheumatism seven years previously,
I had no fear of meeting with any loss of substance
or dislocation. It is of great importance, however,
to bear in mind the fact that there are some cases of
bony anchylosis of the hip joint, and especially those

resulting from the long continuance of strumous disease, in which the head and neck of the thigh-bone are so much destroyed that, for practical purposes, in reference to this operation, the neck of the bone may be said not to exist, although, correctly speaking, it more frequently happens that the head of the bone is destroyed, and the neck, somewhat diminished in size from deficient growth, remains embedded in the acetabulum.

It is a grave pathological error to imagine that because the head and neck of the bone are more or less destroyed in some, this must necessarily be so in all cases of bony anchylosis. This error has, however, been recently committed by Mr. Brodhurst, who, in the course of a correspondence in reference to this operation, has, in the *British Medical Journal* for April 1st, 1871, adduced and figured from St. George's Hospital Museum, 3rd Series, No. 3, a case of bony anchylosis of the hip joint, in which the head and neck of the thigh-bone have been destroyed by strumous disease; and this he assumes to be the ordinary condition of bony anchylosis; for, referring to a diagram of a healthy thigh-bone which I had figured in my paper, merely to show the situation and direction of the subcutaneous division I proposed, Mr. Brodhurst observes—"Thus it will be seen that Mr. Adams made his section, as shown in the diagram, when little or no alteration had taken place in the neck of the thigh-bone. If bony anchylosis had taken place, the section could not have been made as it is described and represented by Mr. Adams, for the sufficient reason that the parts are not there." Again Mr. Brodhurst observes in reply to a letter from

Dr. J. D. Moore, of Lancaster, (see *British Medical Journal*, April 8th, 1871.) " When bony anchylosis has taken place . . . . . the saw has to travel through bone varying in its circumference from eight and a half to six inches; and inasmuch as the neck of the bone is absorbed, and much new bone is deposited, the *section in bony anchylosis can never be made in the manner suggested by Mr. Adams.*" In this letter, Mr. Brodhurst still more positively adheres to his theory that the head and neck of the thigh-bone must necessarily be destroyed in all cases of bony anchylosis of the hip joint ; but for a complete refutation of this theory, it is only necessary to refer to the wood engravings from the eight specimens of bony anchylosis of the hip joint in the Museum of St. Thomas' Hospital, (*See* Plates I, II, III,) which I have appended to this paper, and it will be seen that in five out of the eight cases the neck of the bone is not *absorbed*, as Mr. Brodhurst expresses it, and the operation could have been performed without difficulty. Pathological inquiry, therefore, teaches us that the existence of the head and neck of the thigh-bone, in their natural proportions, is quite compatible with true bony anchylosis of the hip joint; and also in what cases the head and neck of the bone are preserved in their natural proportions, and in what cases the head and neck are destroyed, to a greater or less extent.

In confirmation of the opinion which I have above expressed, that in many cases of bony anchylosis of the hip joint, the head and neck of the thigh-bone remain of their full natural proportions, whilst in other cases they are more or less destroyed—but only in some instances to such an extent as to prevent the

operation of the subcutaneous division of the neck
of the thigh-bone being performed—I need only refer
to the specimens in the various museums of London,
where many typical examples of both these classes of
cases are to be found.

For instance : in the Museum of the Royal College
of Surgeons there are four specimens of bony anchylosis at the hip joint, Nos. 3325, 3326, 3327, and
3327b. In two of these specimens (Nos. 3327 and
3327b), the head and neck of the thigh-bone are of
their full natural size, the anchylosis probably being
the result of rheumatism; and in these cases the
division of the neck of the thigh-bone could have
been easily accomplished ; whilst in the other two
specimens, Nos. 3325 and 3326, the head and neck
of the thigh-bone have been completely destroyed,
evidently by strumous disease; and in these cases,
the operation could not have been performed.

In the Museum of St. Thomas' Hospital, there
are eight specimens of bony anchylosis of the hip
joint; and in these, the operation of dividing the
neck of the thigh-bone could be performed in five
cases, Nos. D 51, D 52, D 53, D 53′, and in a new
specimen not in catalogue; but in the remaining
three specimens, Nos. D 48, D 51′, and D 53′, the
destruction of the head and neck of the thigh-bone
has been so considerable as to prevent the possibility
of the operation being performed.

As the eight specimens in this old and valuable
collection may be taken as fairly representing the
condition exhibited in bony anchylosis of the hip joint,
in the different classes of cases in which it occurs,
I thought it would be interesting to have drawings

made of all the specimens, for the purpose of exhibiting these to the members of the Medical Society,
and now have the pleasure of submitting for inspec-
tion this series of drawings, made by Mr. D'Alton,
an artist of well-known reputation, and for the truth-
fulness of the representations, the drawings may be
compared with the specimens in the Museum. Wood engravings have been made from these drawings and are
appended to the present paper. (*See* Plates I, II, III.)

In the Museum of Guy's Hospital, there are twelve
specimens of bony anchylosis of the hip joint, and in
these the operation of dividing the neck of the thigh-
bone could be performed in eight cases, Nos. 1318^{28},
$1318^{32\ 33}$, 1318^{51}, 1318^{53}, 1318^{55}, 1318^{60}, 1318^{65}, and
1318^{70}. But, in the remaining four specimens, Nos.
1318^{40}, 1318^{45}, $1318^{48\ 49}$, and 1318^{50}, the destruction of
the head and neck of the thigh-bone has been so
considerable as to prevent the possibility of the operation being performed.

In the Museum of St. Bartholomew's Hospital,
there are six specimens of bony anchylosis of the hip
joint, and in these the operation of dividing the neck
of the thigh-bone could be performed in four cases,
Nos. (sub-series) B², B⁴, B⁵, and B⁶; but in the remaining two specimens, Nos. (sub-series) B¹, and
another specimen, B⁶³, not described in catalogue, the
destruction of the head and neck of the thigh-bone
has been so considerable as to prevent the possibility
of the operation being performed.

In the Museum of St. George's Hospital, there are
three specimens of bony anchylosis of the hip joint,
and in these the operation of dividing the neck of the
thigh-bone could be performed in one case, 3rd Series,

Nos. 106 and 107; but in the remaining two specimens (3rd Series, No. 3 and No. 3a), the destruction
of the head and neck of the thigh-bone has been so
considerable as to prevent the possibility of the
operation being performed.

In the Museum of the Royal Free Hospital, there
is one specimen of bony anchylosis of the hip joint,

FIG 9.

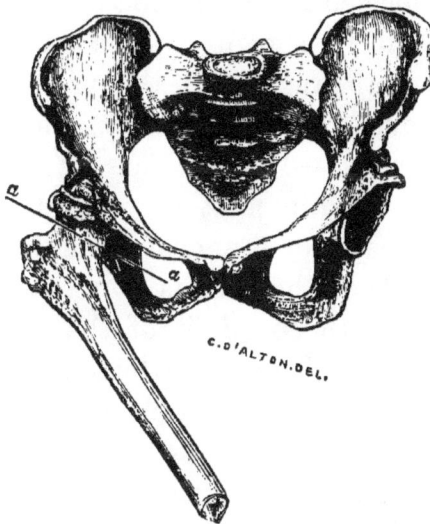

Bony Anchylosis of the right hip joint, with the thigh in a flexed and adducted position.
a a Situation and direction of subcutaneous division of neck of thigh-bone proposed by Mr. Adams.

which, through the kindness of Mr. Gant, I have now
the opportunity of exhibiting to the Society. The
anchylosis probably resulted after the arrest of
strumous disease in the early stage, and the head
of the bone has been only partially destroyed, but the
shaft of the femur is flexed at an angle of 45°, and
adducted to an extent which would cause great lameness
and inconvenience, more especially as the subject was
a female. In this case it will be seen that the division
of the neck of the thigh-bone could have been

accomplished in the direction of the line *a a* without difficulty, although it is an undoubted example of true bony anchylosis. The neck is slightly shortened, and its diameter in the line of incision measures one inch and a half, and its circumference four inches and a half.

Thus it appears that, out of thirty-four specimens referred to, division of the neck of the thigh-bone could be performed in twenty-one cases.

From the facts shown by the specimens above referred to, with regard to the neck of the thigh-bone, it becomes of great practical importance to be able to diagnose:—1. The class of cases of bony anchylosis of the hip joint in which the neck of the thigh-bone remains of its normal length; 2. Those in which the neck of the bone is shortened, but remains of sufficient length to admit of the operation being performed; and 3. Those in which the neck has been so far destroyed as to prevent the operation being performed. There can be no doubt that, in a large proportion of cases, this diagnosis can be made with absolute certainty, and must be based upon the nature of the disease, or morbid conditions producing the anchylosis—viz., whether rheumatic, pyæmic, or traumatic inflammation; or whether it is the result of strumous disease of the joint.

Now, in reference to these points, the following are the conclusions at which I have arrived.

1. In rheumatic anchylosis, no destruction of bone ever exists, and the head and neck of the thigh-bone, therefore, always remain of their full natural size.

2. In anchyolsis after pyæmic inflammation, more especially in its subacute form, from which the patient

frequently recovers, destruction of bone rarely if ever exists, the soft structures only being destroyed.

3. In anchylosis after traumatic inflammation in healthy adults, such as that which occurs after wounds of the joints, and gun-shot wounds in the neighbourhood of the joints, when the joint itself has escaped injury; and in some cases of anchylosis, chiefly from long-retained position, no destruction of bone occurs, as a general rule, even after acute suppurative inflammation, the soft tissues only being involved.

4. In anchylosis after strumous disease of the joint, when arrested in the early stage, without the occurrence of suppuration, or at least of abscess bursting externally, there is generally only a superficial caries of the head of the bone; and the destruction being thus limited in extent, the neck of the thigh-bone remains of its natural length, although practically somewhat shortened by being depressed, or sunk into the acetabulum. In this class of cases, however, the operation can generally be performed.

5. In anchylosis following the more severe forms of strumous disease, in which there has been evidence of caries and necrosis of bone, with abscesses bursting externally, and remaining open a considerable time, generally giving exit to small particles of bone, destruction of the head and neck of the thigh-bone, to a greater or less extent, may be diagnosed; and in all such cases, the operation cannot be performed.

Thus it will be seen that, out of the five classes of bony anchylosis above described, in three classes the head and neck of the thigh-bone remain of their full natural proportions. In the fourth class, although some difficulty may occasionally be met with, the operation can generally be performed ; and it is only in the fifth class of cases that the operation is decidedly negatived.

PUBLISHED AND UNPUBLISHED CASES OF ANCHYLOSIS OF THE HIP JOINT, IN WHICH SUBCUTANEOUS DIVISION OF THE NECK OF THE THIGH-BONE HAS BEEN PERFORMED.

SIX CASES OF BONY ANCHYLOSIS, SUCCESSFUL, AND ONE OF FIBROUS ANCHYLOSIS UNSUCCESSFUL, BEING ALL THE CASES KNOWN TO THE AUTHOR UP TO THE PRESENT TIME, JULY 20th, 1871.

CASE I.

Operation performed by Mr. Adams, December 1st, 1869, and detailed in the present paper. (Successful.)

CASE II.

Operation performed by Mr. J. R. Jessop of Leeds, August 26, 1870. (Successful). *

Margaret R., aged 22, came to the Infirmary on the 1st of August, 1870, seeking relief of the deformity and great inconvenience attendant upon complete fixity of the right hip joint, in such a position that the femur formed with the trunk an angle somewhat more open than a right angle. She walked in a crouching attitude, preferring that mode of progression to the employment of crutches. Her complaint was stated to have originated in an attack of rheumatic fever three years ago, in which the main suffering had fallen upon the right hip and knee. All the

* The following account with the details of this case appeared in the *British Medical Journal*, January 14th, 1871.

joints had recovered completely with the exception of the right hip, which having gone through a long process of medical treatment, including repeated blisterings, had at length become freed of active disease, but firmly fastened in the position mentioned above.

In the hope that the anchylosis might prove to be due to fibrous bands of adhesion, which would yield to forcible extension, Mr. Jessop placed the woman fully under the influence of chloroform, and, with the help of assistants, applied as much weight upon the pelvis and femur as it seemed likely the latter would bear without breaking. When it became clear that the case was one of firm bony anchylosis, it was determined to adopt the ingenious method of treatment recently devised by Mr. W. Adams.

Accordingly, on the 25th August — having previously, in the absence of any published description of the details of the operation, experimented upon the dead subject, with the view of arriving at the best mode of reaching and acting upon the part to be divided—Mr. Jessop proceeded to saw through the neck of the femur. The long narrow knife, which, with the saw, had been devised for the purpose by Mr. Adams, was introduced at a spot three-quarters of an inch behind the great trochanter, and at an equal distance below the posterior superior angle of the same process, and was pushed straight forwards until its point impinged upon the neck of the femur, near its upper border, immediately behind the head. The point of the knife was then carried forward over the neck, so as to clear a passage for the saw, which, with the exercise of some force, was introduced by the side of the knife, until its serrated blade could be

felt to rest upon the bared bone. The knife was
then withdrawn, and the sawing process commenced.
In from ten to twelve minutes, when from the direction of the handle of the saw it was thought that the
blade must have all but completed the severance, an
attempt to extend the limb was made, and the few
remaining fibres of the bone readily gave way with an
audible snap. The limb could now be moved with
the utmost freedom in all directions, and a rough
grating crepitus was distinctly heard as well as felt
at each movement.* During the operation, a few
drachms of dark venous blood escaped from the puncture. The patient was simply put to bed without
any mechanical restraint upon the limb. At each
subsequent visit, she complained of some pain both
in the hip and knee.

On September 8th — a fortnight having elapsed
since the operation—free passive motion of the joint
was commenced; and this was repeated daily, sometimes twice daily, during her entire stay in the
Hospital; she was, moreover, encouraged to move the

* It may here be stated that Professor Lister's rules for accomplishing the exclusion of septic germs were fully carried out. The hands
of the operator, the skin of the part to be operated upon, and the instruments to be employed, were all carefully smeared with carbolised
oil (1 in 5) Thoughout the entire operation, a very efficient supply of
a saturated watery solution of carbolic acid was directed upon the
wound from two of Dr. Richardson's ether-spray producers, and immediately upon the completion of the operation, the wound and neighbouring parts were enveloped in carbolic lac plaster, which was secured
in its position by means of a pad of cotton wool and a roller. On the
following day, and again on the 27th of August, the dressing was
renewed under the protection of the spray. When examined on the
28th, at the expiration of the third day after the operation, the wound
was found to have quite healed; and as neither swelling nor infiltration
of tissue could be discovered, all dressing was discontinued.

leg frequently and freely. During any movement of
the joint she stated, if asked, that she had pain in
both hip and knee; but there was nothing to indicate
that this was at all severe. On September 24th, she
was ordered to move about the ward on crutches,
and frequently to swing the limb backwards and
forwards; and as, when standing upright, the heel
was raised, the anterior part of the foot only reaching
the floor, a thick-soled and high-heeled boot was
ordered for her.

On October 21st, she was sent to the Convalescent
Hospital at Cookridge for three weeks. At this time
she could extend and flex the limb readily, but still
complained of pain in both hip and knee during any
movement. She was able to place the foot on the
floor, but could not bear any appreciable weight upon
it. Crepitus was still distinct. Through the kindness of Mr. Seaton, the honorary surgeon to the
Cookridge Hospital, daily movements of the joint
were faithfully continued. When she returned to the
Infirmary on the 11th November, a marked improvement had taken place. The range of motion in the
new joint was not diminished, and her command over
the limb had increased. She could now place the foot
somewhat firmly on the ground, and could distinctly
bear a little weight upon it. The attempt to use it
in progression gave rise to some pain, but this did
not prevent her from persevering. The crepitus was
now less grating, and gave one the impression that
the surfaces in contact were neither rough nor hard.

During the remainder of her stay in the Infirmary,
which terminated on the 6th December, nearly fifteen
weeks from the time of operation, the improvement

continued. She became able to get along from bed
to bed, without the use of a crutch or stick, in a half
hop, half walk, and quite erect. It was calculated
that she was able to bear from one-fourth to one-
third of her weight upon the limb. By measurement,
the femur was an inch and a half shorter than the
sound one.

As to the result of this case, it is satisfactory to be
able to report that Mr. Jessop, in a letter to Mr.
Adams, dated July 26, 1871, observes: " I have
to-day shewn my two cases at Bradford. My first
case can now walk without any assistance; but as
yet there is a decided limp, and after exercise some
pain. Both limp and pain are, however, diminishing;
and I feel little doubt that after the lapse of another
year, the woman will walk with a firm and natural
tread. A very remarkable change has taken place
in the length of the limb. Some months ago, when
measured, the thigh was an inch shorter than the
sound one. There is now scarcely any perceptible
difference; and the woman had evidently felt this,
for, of her own accord, she informed us that the leg
had increased in length. As she walks, both feet are
on a level, and her boots are properly matched."

The above statement, however, is based upon careful measurement, by our own resident Medical Officer,
Mr. McGill. The movements at the new joints are
remarkably free, and have latterly rather increased
in range.

CASE III.

Operation performed by M. Furneaux Jordan, of Bir-mingham, November, 1870. (Successful).

This case has not yet been published in full, but the following short account of it sent by Mr. Jordan appeared in the *British Medical Journal*, Dec. 24th, 1870. " Emma H., æt. 16, from Wales, had had hip-disease for six years. She had had several sinuses opening and closing during that time. On admission into hospital there was a little oozing from one near the perinæum. The thigh was flexed at right angles to the trunk; and there was unmistakable osseous anchylosis of the hip joint, as revealed by examination under chloroform. I divided the neck of the femur by the method and with the instruments devised by Mr. Adams. The sudden mobility of the joint when the section of the femur was completed was very striking. The adductor longus, and long head of the rectus femoris required tenotomy. The limb was then put into a position which promises a very useful result. It is now three weeks since the operation, and the progress has been most favorable.

" In this case the naturally short femoral neck of early life, made shorter by caries and anchylosis, combined with a very fat gluteal region (since anchylosis the patient has become very stout) required more than ordinary care in every step of the operation."

In a letter to Mr. Adams, dated March 19th, 1871, Mr. Jordan observes, " My case was a marked success; a few drops of moisture came from the wound for a time, but no actual pus. If you publish immediately, add this fact to those in *British Medical Journal*."

CASE IV.

Operation performed by Mr. F. W. Jowers, of Brighton, December 5th, 1870. (Successful).

This case has not yet been published, but the patient was seen on several occasions by Mr. Adams, with Mr. Jowers, while she was in the Brighton Hospital, and also five months after the operation when she was in the Invalid Home at Brighton. The following notes were made at the time of examining the patient, and with Mr. Jowers' assistance.

Anchylosis of both hip joints with both thighs in a flexed position, and useless malposition of the limbs after rheumatic fever four years previously—

Mary Ann Sack, æt. 21, admitted into the Brighton Hospital, under the care of Mr. Jowers, in consequence of anchylosis of both hip joints with useless malposition of the limbs after rheumatic fever four years previously. Both thighs were flexed upon the pelvis at a right angle or more, and when examined under chloroform no motion whatever could be detected at the right hip joint, which was therefore considered to be bony anchylosis, but the left hip joint gave way after considerable force had been employed, and from the sensation conveyed to the hand of the operator on moving the joint, Mr. Jowers considered that the cartilages of this articulation had been destroyed, and that incomplete anchylosis had therefore been broken through. Very severe inflammation followed this operation of forcible extension, but the limb was brought into a straight position, and the case after a time proceeded favourably.

On the right hip joint Mr. Jowers decided to divide the neck of the thigh-bone subcutaneously, according to the method proposed by Mr. Adams, and this operation was performed on the 5th December, 1870. No difficulty was experienced in making the division of the bone, and the small wound made by the tenotomy knife was covered over with cotton wool soaked in carbolised oil. No local redness, swelling, or pain occurred, and the first dressing was left undisturbed for eighteen days. The wound was then found to be quite healed, and no suppuration whatever had occurred.

Immediately after the operation the limb was put up with a long outside splint, and a straight well padded splint was also applied to the back of the limb, and carried above the pelvis, so that the limb was kept in a perfectly straight line with the body, the back splint appearing to be a very useful addition.

On the 4th January, 1871, the long splint was removed, and a weight of two pounds attached to the leg substituted. Passive motion was also commenced.

On the 11th January she was allowed to get up, and on the 26th April she left the Brighton Hospital.

On the 7th May, 1871, Mr. Adams saw this patient with Mr. Jowers in the 'Invalid Home,' at Kemp Town, Brighton. Both the legs were in a perfectly straight position, and she was able to walk about the room without any assistance whatever. No motion could be traced at the right hip, so that firm osseous union seemed to have occurred at the seat of operation. In the left hip-joint it was doubtful whether any motion existed, but Mr. Jowers thought there was not complete anchylosis. Notwithstanding this condition of both the hip-joints, this patient was enabled to sit upright at meal-times on a wooden bench, sitting down upon it, and getting up again without assistance. This was evidently due to a preternatural mobility which had taken place to a remarkable extent at the articulations of the lumbar vertebræ, as a compensative effort of nature.

CASE V.

Operation performed by Mr. John Croft at St. Thomas' Hospital, March 4th, 1871. (Unsuccessful).

This case, which was one of fibrous anchylosis in a child, æt. 8 years, has not yet been published by Mr. Croft, but in the paper "On the Selection of Cases, &c.," read by the author at the Medical Society of London, and published in the present pamphlet, some account of this case has been given by Mr. Adams, who was present at the operation performed by Mr. Croft, see page 19 of the present pamphlet.

CASE VI.

Operation (second case) performed by Mr. T. R. Jessop, of Leeds, March 16th, 1871. (Successful).

The details of this case have not yet been published, but in a letter to Mr. Adams, dated March 16th, 1871, Mr. Jessop writes as follows :—" I have to-day again divided the neck of the femur by your subcutaneous method. The patient, a woman, æt. about 30, has had her right hip joint anchylosed at *less* than a right angle ever since she was 10 years old—the result apparently of strumous hip disease. The union was bony. The whole limb is wasted, or rather undeveloped, but is of size and strength sufficient to be of much use when straightened. I found the neck shortened, thickened and much harder than natural. Considerable patience and power were needed in the operation, which however was satisfactorily accomplished. The sawing occupied nearly a quarter of an hour."

In a letter to Mr. Adams, dated July 26th, 1871, Mr. Jessop, observes :—

" I append a few facts relating to my second case.

" Martha Fearnley, aged 27, admitted March 6th, 1871. When six years old she fell upon the right hip, thereby starting the ordinary disease of that joint. For two years she limped about in pain, and then for twelve months was kept in bed under treatment. It was not, however, until she reached the age of 15, that the disease could be said to have become arrested, and then the knee was drawn up so that she could not touch the ground with her toes when standing erect. For some years the thigh continued to bend more and more upon the body, and finally became firmly fixed at considerably less than a right angle. During this time no abscesses ever burst externally. On admission, the joint was firmly anchylosed at less than a right angle, and the thigh was adducted so that the knee crossed the opposite limb.

"*March 9th.*—Chloroform was administered, and forcible attempts were made to straighten the limb; but without any effect. All present concurred in the belief that bony union had taken place.

"*March 16th.*—Mr. Adams' operation was performed. The bone was found to be very hard and much thickened. The sawing occupied thirteen minutes. After the completion of the operation some force was needed in straightening the limb to overcome the resistance of contracted muscles. The antiseptic treatment was adopted during and after the operation. The limb was kept extended by means of a weight attached to it, and suspended over a pulley.

"*March 21st.*—Wound quite healed.

"*April 4th.*—Weight removed, and passive motion commenced; this causes *extreme* pain. During the next few days an occasional attempt was made to move the limb, but the pain was so intense that it was impossible to persevere.

"*April 19th.*—Chloroform was administered, and then the limb was moved *very freely* in all directions for some minutes. After this, daily movements were persevered with, causing much though lessening suffering.

"*May 1st.*—Sat up for an hour out of bed. On standing upright the toes of the side operated on just touch the floor. She can swing the limb to and fro.

"*May 20th.*—She gets up daily, and walks about with crutches, swinging the leg freely as she moves. As yet no weight can be borne on the leg. She left the Infirmary about the middle of June.

"*Condition on July 26th.*—The limb is considerably *longer ;* she can now fairly reach the floor with the sole of the foot. She can bear fully one-half her weight upon the limb, can swing it freely about without pain, and uses one crutch only. She states that every week she can see distinct progress, and is very sanguine as to the ultimate result.

" I have little doubt this case will ultimately equal the other."

Note.—The success which has attended the efforts to obtain a new joint in Mr. Jessop's cases, will in future induce me to follow the method he has adopted in the after treatment, and persevere in moving the limb notwithstanding a certain degree of pain, and frequently moving it under chloroform.—W. A.

CASE VII.

Operation performed by Mr. James Hardie of Manchester, June 7th, 1871. (Successful).

The details of this case have not been published; but in a letter to Mr. Adams, dated July 12th, 1871,

Mr. Hardie writes as follows :—"You will no doubt be pleased to hear of another successful case of your new operation for anchylosed hip joint. A case of rheumatic anchylosis of three years' standing came in my way some time ago. The hip was flexed to an angle of 100°, and the patient (a female, æt. 23) was quite unable to walk further than a quarter of a mile at a time without support. Her gait was of course extremely awkward. I operated five weeks ago, and after dividing the adductor longus and rectus, the limb became perfectly straight. I tried to obtain the formation of a new joint, but fear that will not be succeeded in. There was not a single complaint made by the patient after the operation, and the puncture healed in a fortnight.

"I experienced considerable difficulty in accomplishing the section of the bone. It seemed more a process of filing than of sawing, and the saw got locked very frequently. It occurred to me that the back of it might be advantageously thinned down a little more. I shall report the case by and bye."

SUBCUTANEOUS OSTEOTOMY.

NOTES ON THE FOREIGN LITERATURE OF SUBCUTANEOUS
OSTEOTOMY, BY HENRY DICK, M.D.

Jules Guérin* must be regarded as the first who divided bones subcutaneously. In page 34 of the Treatise referred to, he observes, in speaking of the different affections in which subcutaneous operations may be applicable, " 5th. The removal of a small exostosis on the superior and anterior part of the tibia, followed by absorption of the *débris* of the tumour, without any consecutive symptoms of inflammation."

Guérin† was the first who operated subcutaneously in joints, as in the removal of loose cartilages from the knee-joint. Syme claimed to be the first, but could not make good his claim against Guérin.

Langenbeck,‡ in the Schleswig-Holstein war of 1848, first introduced the use of small straight pointed saws for section of bones, and remarks, that after the introduction of these instruments, the idea

* " Essais sur la Méthode sous-cutanée," page 34. Paris, 1841.
† " Mémoire sur les Plaies sous-cutanées des Articulations," page 75.
‡ " Subcutane Osteotomie," von B. Langenbeck, M.D. Berlin, 1854.

of subcutaneous resection of bones dawned upon
him. In fact, several resections were made which
resemble subcutaneous operations, or merit the name
of them. Langenbeck removed in this way (with a
small and pointed saw) exposed portions of bones
when suppuration existed, introducing the subcutaneous saw through the narrow opening in the
upper arm in bullet wounds, without further disturb-
ing the soft parts. In this way he made resections
of small portions of bones. He then decided to
employ this method for other diseases of bone.

In his clinique, in the winter of 1852, he practised
the operation in pseudarthrosis femoris (fibrous
anchylosis). He further operated on bony anchylosis
of the knee where he had previously tried to break
them under chloroform.

These two cases of bony anchylosis occurred in two
young and healthy individuals; one the result of
penetrating wounds, the other after rheumatism.

Langenbeck describes his instruments as follows :

1. A drill in the shape of a gouge $\frac{1}{4}$-inch in
diameter, or what is called in England a grill-bit,
placed in a brad or click, very similar in its use and
construction to that employed by cabinet makers.

2. A narrow pointed saw, $\frac{1}{8}$th of an inch wide, and
four inches long.

3. A strong incision knife, such as is generally
used for resections.

Langenbeck made his first subcutaneous section of
bones on ricketty bones in the tibia and fibula. His
modus operandi in these cases was as follows. He
made a small incision, $\frac{1}{2}$ to $\frac{3}{4}$ of an inch in length, on
the inner surface of the tibia, at a right angle with the

longitudinal axis of the tibia, separating the skin and periosteum at one cut. The drill was then applied, and pierced the tibia in an oblique direction. That the drill had pierced the bone completely, and arrived in the space between the fibula and tibia, was proved by the lost resistance, so there was no danger of injuring the posterior tibial artery or nerve. The small saw was then introduced into the hole made by the drill, and the bone sawn through subcutaneously in an oblique direction, corresponding to the wound in the skin, until the thin bridge, or the shell, or compact tissue of the bone was reached.

The thickest part of the bone now being sawn through, the rest was easily broken, and the straightening of the bone could be left until the inflammation and suppuration produced by osteotomy had ceased, and Langenbeck was in favour of straightening the bone at once. Suppuration only lasted in the first case four or five days; but in the second, fourteen days.

Langenbeck thinks that badly healed fractures (obliquely healed), or severe ricketty deformities, or anchylosis are cases for subcutaneous osteotomy.

Langenbeck's Cases of Subcutaneous Operations of Bones.

Langenbeck gives three cases :—

1. Severe deformity of the tibia, which was subcutaneously cut, as above described (successful).

2. Also a ricketty case of tibia.

3. Deformity after fracture of the tibia and fibula, at the age of five years. The patient was thirty-five years of age when operated upon. The ends of both

fractured bones seemed to be united in one callus, and great deformity existed.

Langenbeck observes (in his pamphlet) that until then he had never performed subcutaneous section in real bony anchylosis of joints. He had only done so once, in a pseudarthrosis femoris (fibrous anchylosis).

Langenbeck is opposed to Rhea Barton's operation. Langenbeck proposed subcutaneous operation of bones in anchylosis, but had only performed it in ricketty bones.

In two cases of anchylosis of the knee-joint, one after a wound, the other after rheumatism, the patients were willing to have the operation performed if Langenbeck had assured them there would be no danger in the operation, which he refused to do, as he believed there might be some danger of pyæmia, as in all other operations ; and the patients left the hospital without being operated upon.

Langenbeck's conclusion on the operation are :—

1. Bones can be cut subcutaneously like tendons, muscles, &c.

2. Bones should only be partially cut subcutaneously, the partial section being preferable to the complete division.

3. Healing of subcutaneous section of bone will not take place as in simple fracture, for the reason that the drill and saw produce some powder, which acts like a foreign body, and therefore some suppuration will take place.

Subcutaneous section is indicated in deformity of

bones, and bony anchylosis, and where we cannot rupture, the bones being too solid.

Langenbeck gives three cases of subcutaneous osteotomy, but they are all for rickets of the tibia.

Meyer did not make the section of bone subcutaneously, but by a small opening, approaching to the subcutaneous section.

Wernher,* speaking of Meyer's osteotomy, observes, that Meyer only practised it for ricketty deformities. But he (Wernher) thinks this method may also be employed for bony anchylosis of the joints. Meyer's operation cannot strictly be called subcutaneous, since the bone was exposed by the osteotome, and divided either by the knife-saw, chain-saw, or the trephine.

Gross,* in his System of Surgery, states that Pencoast of the Jefferson Medical College, in the winter of 1859, performed the following operation:—"It consisted in perforating with a stout gimlet the femur subcutaneously through a single opening at half-a-dozen points, just above the knee, and then forcibly breaking the bone.

"The limb was then placed in appropriate apparatus, the upper end of the inferior fragment forming an angle with the upper fragment projecting into the ham. A large abscess formed at the seat of fracture, but with this exception the case progressed favourably, and the boy made a good recovery, the foot coming down well, and the limb being as nearly straight as

* See "Handbuch der allgemeinen und speziellen Chirurgie," von D. A. Wernher, Giessen, 1855, page 745.
† "System of Surgery," Philadelphia, 1862, page 85.

could be desired, in such a condition of the knee-joint."

Gross* states that Professor Brainard, of Rush College, in 1860, performed a similar operation, but in this case the femur was divided through its condyles by means of the peculiar perforator employed by that surgeon, the use of which is less liable to be followed by severe inflammation than that of the gimlet. The patient recovered with a good limb.

In another case of anchylosis of the knee, the patella was detached subcutaneously from the femur and tibia, and excellent motion at the joint obtained.

Observations by Dr. Henry Dick.

Subcutaneous osteotomy was first performed in this country by Mr. L. Stromeyer Little, on an anchylosed knee, but his operation differed considerably from Mr. Adams'.

Subcutaneous osteotomy on the neck of the femur was first performed in this country by Mr. William Adams for anchylosed thigh. His instruments were a small saw, and a long tenotome.

There is no doubt that previous to this, subcutaneous osteotomy had been performed both in Germany and America, but still Mr. W. Adams' mode of procedure differed considerably, and, after searching, as far as in my power, I believe that Mr. William Adams was the first who performed the subcutaneous section of the neck of the thigh-bone. All the subcutaneous operations on bone were either performed

---

* Gross, *op. cit.*

on long bones, or on the anchylosed knee, and the wounds generally made were too large to merit the name of subcutaneous operations. Mr. Adams' operative procedure with the small instruments employed alone merits the name of "subcutaneous operation."

ABSTRACT OF A PAPER BY MR. L. STROMEYER LITTLE, READ AT THE ROYAL MEDICAL AND CHIRURGICAL SOCIETY, MAY 23RD, 1871. COPIED FROM THE "LANCET," JUNE 3RD, 1871.

"A case of Bony Anchylosis of the Knee Joint treated by Subcutaneous Section of the Bone.

"The author in this paper gives an account of a case of bony anchylosis of the knee-joint in a child aged fourteen, in whom the limb was fixed at a right angle. The anchylosis was divided subcutaneously by means of a carpenter's chisel, and by an extending apparatus the limb was straightened so as to allow of locomotion three weeks after the operation. The author discusses the plan of dividing the long bones by means of a saw for the cure of deformity, and concludes that for bony anchylosis of the knee joint subcutaneous osteotomy by means of a saw is impracticable. The case is believed to be the first instance where subcutaneous osteotomy has been performed in this country for the relief of bony anchylosis of a large joint.

E

"The President, having assisted at the operation performed by Mr. Little, was able to bear testimony to the fidelity of the description of it in the paper, and to its success in restoring a useful limb. Had Mr. William Adams been present, the President would have inquired whether he had heard of Mr. Little's operation on the knee before performing, a year afterwards, a similar operation on the hip joint.

BOSTON MEDICAL
JAN 8 1904
LIBRARY

A LETTER FROM MR. ADAMS, REPRINTED FROM THE "LANCET," JUNE 10TH, 1871.

Bony Anchylosis of the Knee Joint treated by Subcutaneous Section of the Bone.

SIR,—Referring to the report in your last impression of a paper with the above title by L. S. Little, F.R.C.S., read at the last Meeting of the Medical and Chirurgical Society, I should be obliged if you would allow me to answer a question which the President, Mr. Curling, stated he wished to ask me had I been present.

Your report states : " Had Mr. William Adams been present, the President would have inquired whether he had heard of Mr. Little's operation on the knee before performing, a year afterwards, a similar operation on the hip joint."

I regret having been unavoidably absent at the Meeting, or I should have been only too glad to have borne testimony to the practical interest and value of the operation performed by Mr. L. S. Little, and described in the paper.

As an answer to the President's question, I may state that I had been informed of Mr. Little's operation by our mutual friend, Dr. Henry Dick ; and it was in a conversation with him, and very much at his suggestion, that I first proposed simply to divide the neck of the thigh-bone subcutaneously by a very small saw, instead of using a chisel, to cut through part of the bone, and afterwards break through the remainder by forcible extension, as in the method employed by Mr. Little, and referred to by Dr. Little in "Holmes' System of Surgery," second edition, vol. iii., p. 722, at least presuming this to be the same case as that described at the Medical and Chirurgical Society.

Whilst Mr. Little's operation was admirably adapted to the knee joint, I thought that an operation attended with much less violence, and of a more perfectly subcutaneous character, might be performed at the hip joint. The difficulties in the construction of a saw were overcome, after various experiments, by Mr. Blaise; but, still feeling doubtful of the practicability of the operation I proposed, in the event of failure I was fully prepared to adopt the method employed by Mr. Little, and took with me long narrow chisels with oblique cutting edges for this purpose.

In the course of my operation on the neck of the thigh-bone, I found no difficulty in making a simple division by means of the small saw I employed; and therefore it will at once be seen that, in point of practical details, as well as in the essential principle of avoiding all mechanical force in breaking through bone, the operation performed by myself on the hip joint was essentially different from that performed by Mr. Little on the knee joint. Indeed, as stated in your report, the author of the paper observed, "for bony anchylosis of the knee joint subcutaneous osteotomy by means of a saw is impracticable."

In one important respect, however, both operations were similar —i. e., they were both of a subcutaneous character, and, so far as I know, the credit of having first performed a subcutaneous operation of this character in England belongs to Mr. Little, though subcutaneous osteotomy, with various modifications in the operations performed, had been previously practised by Langenbeck, Gross, Bauer, Meyer, and others.

I am, Sir, yours, &c.
WILLIAM ADAMS.

Henrietta Street, June 6th, 1871.

QUOTATION FROM AN ARTICLE ON ORTHOPÆDIC SURGERY, BY
DR. LITTLE IN HOLMES' "SYSTEM OF SURGERY," REFERRING
TO THE OPERATION DESCRIBED IN MR. L. S. LITTLE'S
PAPER.

*" Forcible subcutaneous separation of the tibia and
femur in true bony anchylosis of the knee joint."**

"This operation originally proposed, and carried
out successfully by Langenbeck and Gross of Philadelphia, has been performed in this country with some
modifications by Mr. Little at the London Hospital.
The operation consists in making a small incision in
the integuments and fibrous tissues at the side of the
articulation, parallel to the plane of the natural
articulating surface of the tibia. The length of this
incision should correspond with the width of a narrow
sharp cutting, well tempered ordinary chisel, say two
to three lines in width, which being driven in different
directions between the ends of the femur and tibia,
united by osseous material, so effectually weakens the
connection between the adherent surfaces, that
straightening and bending of the limb can, with the
exercise of "gentle violence" with the hands, be
readily effected. When the surgeon remembers that
the joint has been destroyed by the diseased process
which produced the bony anchylosis, he will not be
surprised to learn that this surgical subcutaneous
chiselling asunder of bones is not followed by any of
the serious consequences known to follow wounds of

* "A System of Surgery," by various Authors, edited by Holmes.
Second Edition, Vol. III, page 722. London 1870.

the joint, and that with or without simultaneous sections of knee tendons, as may appear requisite, the limb may be placed in the desired curative position. It may seem superfluous to remark that the operation is perfectly safe in the hands of the surgeon who avoids injury of the important nerves and vessels about the articulation. It cannot fail to become a standard operation for relief of knees affected with *bony* anchylosis in a bent position, and is in every respect infinitely preferable to the operation of knee resection when it has been performed for mere anchylosis."

APPENDIX.

NO. 1.   LETTER FROM MR. ADAMS.

(Reprinted from the British Medical Journal, February 18, 1871.)

THE SUBCUTANEOUS DIVISION OF THE NECK OF THE THIGH-BONE.

SIR,—In the paper read by me at the meeting of the British Medical Association at Newcastle on August 10, 1870, and published in the *British Medical Journal* on December 24th, 1870, I described what I believed to be a perfectly new and original operation, by which, in certain cases of bony anchylosis of the hip joint, with the limb in a deformed position, the neck of the thigh-bone might be divided subcutaneously by means of a small saw, a quarter of an inch in width, and having a cutting edge a quarter of an inch in length, introduced through a punctured wound made with an enlarged tenotomy knife a little above the great trochanter, and carried directly down to about the centre of the neck of the bone. By this operation, which I performed at the Great Northern Hospital on December 1st, 1869, cases of extreme deformity may be immediately rectified, and the limb brought into a straight position.

I was not then aware that any such operation had ever been suggested or performed by any one; but in a work on *Deformities*, published within the last few weeks by Mr. Brodhurst, and which is essentially a reprint of lectures previously published by him in the *Lancet*, he has claimed to have performed an operation *of this character* in the year 1865.

In the work alluded to, Mr. Brodhurst describes four operations, as applicable to cases of bony anchylosis, and at page 152 adverts to the fourth operation in the following terms.

"4. Where it is not desired to obtain motion, but only to rectify a false position of the limb, the bone may be divided subcutaneously, and an improved position may be given. I performed an operation *of this character*, with the assistance of Dr. Richard Brown and Mr. Potter, in the year 1865, and have subsequently had occasion to repeat it. In the present year Mr. W. Adams has, at the Great Northern Hospital, also in a similar way, cut through the neck of the thigh-bone."

No details whatever are given by Mr. Brodhurst, nor is it even stated that the operation was performed by him at the hip joint. Moreover, it is a remarkable fact that in the original lecture, published in the *Lancet*, no mention is made of this operation having been performed either by himself, or by any other surgeon.

The following quotation is from the lecture as published in the *Lancet*, February 20th, 1869.

"*The Treatment of Bony Anchylosis.*—There are three operations which may, under certain circumstances, be done, to restore motion, or to improve the position of the limb—viz., 1st, to remove a wedge of bone; 2nd, to break through the anchylosis, after drilling through the new bony formation; 3rd, to make a false joint."

In Mr. Brodhurst's work on *Deformities*, however, just published, and stated in the preface to be a re-issue of these lectures, the subcutaneous division of the bone is mentioned as a fourth operation for anchylosis in the terms of the quotation above given.

Thus, it appears that in his lectures, published in 1869, Mr. Brodhurst mentions only three operations for the treatment of bony anchylosis; and in his work, published in 1871, he mentions four operations for the same class of cases; the fourth being that which I had in the interval published at the Newcastle meeting in August, 1870, and of which I claimed to be the originator.

Now, I can only explain the singular omission of all notice of any such operation in his published lectures in 1869 by assuming that Mr. Brodhurst had either forgotten that he had ever performed such an operation, or that he attached so little importance to it as to omit all mention of it as a surgical procedure.

In the absence of any published record of Mr. Brodhurst's case, or cases, I could have no knowledge of any such operation having been performed; but now I hope we shall be favoured in the *Journal* with all the necessary details as to the nature of the case; the joint operated upon, since this is not mentioned; the mode of performing the operation, and instruments used; by whom and when made; the progress of the case as to the occurrence of suppuration, or not; and the result of the case, which would be of additional interest from the length of time that has elapsed.

I think, Sir, you will agree with me that such details are called for, before I could surrender my claim to originality with respect to the operation in question.

I am, etc.,   W. ADAMS.

Henrietta Street, Cavendish Square, February 13th, 1871.

NO. 2.   LETTER FROM MR. BRODHURST.

Reprinted from the British Medical Journal, February 25th, 1871.

SIR,—My attention has been directed by Mr. William Adams to a letter which he has addressed to you, and which appears in your current number in the following terms. " I think it right to direct your attention to a letter written to the editor of the *British Medical Journal*, and which is printed in the Journal of to-day (February 18th) with respect to a claim you have set up in your recently published work on *Deformities* to an operation of a similar character to that performed by myself, and described as subcutaneous division of the neck of the thigh-bone." And in his letter to you, Mr. Adams states: " In the paper read by me at the meeting of the British Medical Association at Newcastle on August 10th, 1870, and published in the *British Medical Journal* on December 24th, 1870, I described what I believed to be a perfectly new and original operation, by which, in certain cases of bony anchylosis of the hip joint with the limb in a deformed position, the neck of the thigh-bone might be divided subcutaneously by means of a small saw, a quarter of an inch in width, and having a cutting edge a quarter of an inch in length, introduced through a punctured wound made with an enlarged tenotomy knife a little above the great trochanter, and carried directly down to the centre of the neck of the bone. By this operation, which

I performed at the Great Northern Hospital on December 1st, 1869, cases of extreme deformity may be immediately rectified, and the limb brought into a straight position." In my work on the *Deformities of the Human Body*, I treat in Chapter XI. of true anchylosis, and I describe four operations; namely, 1, to remove a wedge of bone; 2, to break through the anchylosis after drilling through the bony formation; 3, to make a false joint: and 4, to divide the bone across subcutaneously, and thus to restore the position of the limb. All these are subcutaneous operations except the first mentioned. This was Barton's operation, which I need not now describe. And, in using the term subcutaneous, as applied to the fourth mentioned operation, I have done so to distinguish it more especially from Barton's.

Subcutaneous osteotomy has been practised and written of by Langenbeck, Gross, Pancoast, Brainard, Bauer, Meyer, Linhardt, and others. Some have practised subcutaneous osteotomy by making the external incision less than half-an-inch in length, while others have preferred a freer external incision. In the case to which I allude, at page 152, I operated in 1865, by making a small external incision, sufficient only to use a very small saw with ease. This operation was not at the hip, but it was done on account of considerable deformity which occasioned lameness. The patient was a Member of the House of Commons and a sportsman, and this deformity prevented him from taking that exercise to which he had been accustomed.

But, in 1861, I cut through the neck of the thigh-bone subcutaneously in a case where bony anchylosis had taken place at the hip joint; and in the following year, I brought the details of the case before the Royal Medical and Chirurgical Society, hoping that they were of sufficient interest and importance to gain a place in the *Transactions* of the Society, as a continuation of another paper on Fibrous Anchylosis which had already been published in the *Transactions*, Vol. XL. This paper, however, was not published, and a slender record alone appeared in the Society's *Proceedings*, Vol. IV. In this instance, the external incision extended to about two and a half inches in length, the limb being contracted at an angle of 45 degrees, and subsequently a suppurating sinus, which, however, was superficial, was extended into the wound.

In my communication to the Society, this operation was treated of as subcutaneous osteotomy, and as a matter of fact, the bone was never exposed to view. " The wound, I state, healed in about its entire extent by the first intention, and in three weeks it was firmly cicatrised, so that passive motion could be freely employed."

In the former operation to which I have alluded, and where a button-hole aperture alone was made, the wound took much longer time to heal. I have now operated in several cases of a similar kind, and I prefer to make such an incision as shall afford ample room for the protection of the soft structures.

With this case—nine years old—before him, I may be allowed, perhaps, to express some surprise that Mr. Adams should describe his operation of December 1st, 1869, as "a perfectly new and original operation."

I am, &c.,
B. E. BRODHURST.

20, Grosvenor Street, February 20th, 1871.

NO. 3.   LETTER FROM MR. ADAMS.

(Reprinted from the "British Medical Journal," March 4, 1871.)

SIR,—In Mr. Brodhurst's letter, published in the *British Medical Journal* of the 25th February, in reply to mine of the 18th February, he has still left unanswered the questions which I put to him as necessary to the establishment of his claim to priority in the operation of subcutaneous division of the neck of the thigh-bone.

Mr. Brodhurst's operation, in the year 1865, on which he rests his claim in his work on *Deformities*, p. 152, was, by inference, assumed by me to have been performed on the hip joint, since it was placed in juxtaposition with my hip joint operation, and used by Mr. Brodhurst in his argument as being *of a similar character*.

As the joint operated upon was not stated, I begged, in my letter of the 18th February, that we might be favoured with this information as well as with other details. Mr. Brodhurst, however, has not given this information, but in his reply states, " this operation was not at the hip, but it was done on account of considerable deformity, which occasioned lameness." I think, however, I can supply your readers with the information which Mr. Brodhurst declines to give, and state that the operation in question was performed on the *great toe* of a gentleman, a portion of the metatarsal bone being removed from what Mr. Brodhurst describes as a considerable deformity. In this region I have no doubt that the "button-hole" aperture was amply large enough to admit of his using "a very small saw" with ease, especially as Mr. Brodhurst, alluding to the relative merits of large and small incisions, remarks, "I have now operated in several cases of a similar kind, and I prefer to make such an incision as shall afford ample room for the protection of the soft structures." Although we are not told either of the precise nature, or the result of this case, we are informed that "the wound took much longer time to heal" than an operation on the hip joint, with which it was compared, so that the process of the case was not in accordance with the results of subcutaneous surgery.

On the above case, on which Mr. Brodhurst has based his claim, I am sure I need make no further comment than to ask any of your readers whether such an operation as this, performed by Mr. Brodhurst on the great toe, can be compared with the subcutaneous division of the neck of the thigh-bone, as performed by myself, or even correctly described as one of a *similar character*.

In Mr. Brodhurst's letter, however, as if fearing that the operation of 1865 might, upon inquiry, break down, he has imported a fresh operation, upon which he now, for the first time, bases his claim to the subcutaneous division of the neck of the thigh-bone as far back as the year 1861.

Mr. Brodhurst observes, " But in 1861, I cut through the neck of the thigh-bone subcutaneously, in a case where bony anchylosis had taken place at the hip joint; and in the following year, I brought the details of the case before the Royal Medical and Chirurgical Society, hoping they were of sufficient interest and importance to gain a place in the Transactions of the Society . . . . This paper, however, was not published, and a slender record alone appeared in the Society's *Proceedings* . . . . In my communication to the Society this operation was treated of as subcutaneous osteotomy."

Let us now see what the operation really was, according to the abstract of the paper published in Vol. IV. of the *Proceedings* referred to (p. 97), premising that such abstracts, as a general rule, are supplied by the author himself. The title of the paper is as follows, "Further Observations on Anchylosis, with an Account of a New

Operation for the Restoration of Motion, when Articular Inflammation has resulted in Synostosis."

The case was one of hip joint disease of many years' duration, occurring in a young lady, æt. 25. Abscesses had formed, and bony anchylosis is said to have ensued with the limb in a deformed position. " It was determined to divide the neck of the thigh-bone to remove the necrosed portions of bone, and to form a false joint. The operation was commenced by making an incision two and a half inches in length, which commenced one inch and a half above the great trochanter, and which was carried downwards, and outwards to the outer side of the trochanter itself. The upper portion of the incision was extended upwards and inwards for two and a half inches, until it fell into a suppurating sinus immediately below Poupart's ligament. The neck of the femur was sawn through, and the sharp edges of the bone were removed, as well as the necrosed bone of the acetabulum."

Neither in the title of the paper, nor in the text, is the word *subcutaneous* even used; nor is any allusion whatever made to subcutaneous osteotomy; neither is any such allusion made in his work on *Deformities*, p. 150, where this case is also alluded to, and placed in juxtaposition with one of Dr. Barton's for establishing a false joint in anchylosis; so that putting this operation in the class of subcutaneous operations has been clearly an after-thought of Mr. Brodhurst's.

Now, Sir, if operations of this magnitude, with large external incisions—in the present case stated to be two and a half inches in length, and then extended to five inches—(from the great trochanter to Poupart's ligament)—admitting the free use of the saw, chisel, and gouge, for the purpose of dividing bone, and removing necrosed bone, are to be called subcutaneous operations, then I must confess myself to be entirely ignorant of the principles and practice of subcutaneous surgery, the very essence of which consists in the small size of the external wound, and the entire exclusion of air from the divided structure, whether tendon, muscle, nerve, or bone; by which means we ensure a more simple and more perfect reparative process than in open wounds as well as freedom from suppuration, with all its attendant dangers.

If Mr. Brodhurst expresses surprise that I should still claim originality for my operation of subcutaneous division of the neck of the thigh-bone, I might well be excused using stronger language in condemning the attempt made by Mr. Brodhurst to appropriate my operation upon such evidence as he has adduced, and which I have now fairly analysed, leaving his claim to be decided by the further criticism and judgment of the profession.

I am, etc.,

WILLIAM ADAMS.

Henrietta Street, Cavendish Square, February 28th, 1871.

NO. 4. LETTER FROM MR. BRODHURST.

Reprinted from the British Medical Journal, March 18th, 1871.

SIR,—In my recently published book I have described four operations for bony anchylosis, three of which are subcutaneous, and one only, namely, the first mentioned, which is known as Barton's operation, is by open wound ; and I refer to two cases, one of which is placed in the third category, as an illustration of subcutaneous osteotomy with the formation of a false joint ; and the other comes into the fourth category, which comprises operations for the rectification of a distorted position of a limb ; and I mention that the former of these operations was done by making in the first instance an external incision to the extent of two and a half inches, and that in the second the external opening was only sufficiently large to allow the use of the smallest saw. The first mentioned operation was performed at the hip in 1861, and the second was done on the first metatarsal bone in 1865.

Passing over various mistatements and exaggerations which Mr. Adams has introduced into his last letter, I will place side by side my paragraph to which reference is made, and that of Mr. Adams, in which he alludes to it, that your readers may judge whether his deduction is fair.

" 4. Where it is not desired to obtain motion, but only to rectify a false position of the limb, the bone may be divided subcutaneously, and an improved position may be given. I performed an operation of this character, with the assistance of Dr. Richard Brown and Mr. Potter, in the year 1865, and have subsequently had occasion to repeat it. In the present year Mr. W. Adams has, at the Great Northern Hospital, also in a similar manner cut through the neck of the thigh-bone." *(Deformities of the Human Body, p. 152).*

The following is Mr. Adams' reference to this statement.

" Mr. Brodhurst's operation in the year 1865, on which he rests his claim in his work, *On Deformities*, p. 152, was by inference assumed by me to have been performed on the hip joint, since it was placed in juxtaposition with my hip joint operation, and used by Mr. Brodhurst in the argument as being of *a similar character*.

Now, sir, I contend that Mr. Adams has no right whatever to assume what he has done with regard to my operation, nor to make the statement that I rest my claim to subcutaneous osteotomy on this operation. If he had never heard of my case of division of the neck of the thighbone, there might be some excuse for his statement; but not only have I talked of it with him, but he evidently knows more than I have told him, for he says that I made " free use of the saw, chisel, and gouge," in performing this operation. This, however, is entirely a mistake. In bringing my case before the first Medical Society in the metropolis, I consider that I did all that was necessary to establish my claim.

Mr. Adams brought his case before various societies—before the Association, for instance—and in his abstract the following occurs.

" The author referred to the various operations which have been proposed and adopted for bony anchylosis of the hip with deformity, such as Rhea Barton's operation, and also that proposed by Louis Sayre of New York, which he had performed in two cases.

Barton's operations was, perhaps, sufficiently distinct from that of " the author " to enable him to mention it. It consisted of a crucial incision over the great trochanter seven inches in length and five inches in a horizontal direction. With a fine saw he then divided the bone transversely between the two trochanters. Judging from his abstract, it would appear that Mr. Adams at that time (it was published September 1870) knew nothing of any other cases.

Proceeding with his abstract, we find that " no inflammation whatever had followed the operation ; and the author, therefore, felt justified in comparing this operation of the subcutaneous division of bone, or subcutaneous osteotomy, with the subcutaneous division of tendons." Clearly, then, this was entirely an original idea. Mr. Adams, having divided tendons subcutaneously, determined to divide bone in the same manner, and, seeking about for a term to describe his operation, he found *subcutaneous osteotomy*. If there is any meaning to be attached to the above description, it is that which I have given to it.

But, unfortunately, Mr. Adams had previously described this same case before another Society—the Medical Society of London—where, instead of the description being " no inflammation whatever had followed the operation," the report is as follows: " On the 7th December" (the operation having been done on the 1st December) " a few drops of pus only escaped from the superficial wound. December 22nd. No febrile symptoms or deep suppuration going on; the superficial wound nearly healed. There never has been more than two or three drops of pus on the lint in the morning. He got up to day for the first time, just three weeks since the operation."

Referring back to Mr. Adams' indignant letter of March 4th, the following occurs relating to my case : " If operations of this magnitude," &c., (for he is shocked at mine of two and a half inches), " are to be called subcutaneous operations, then I must confess myself to be entirely ignorant of the principles and practice of subcutaneous surgery, the very essence of which consists in the small size of the external wound, and the entire seclusion of air from the divided structure, whether tendon, muscle, nerve, or bone, by which means we insure a more simple and more perfect reparative process than in open wounds, as well as freedom from suppuration, with all its attendant dangers."

It is really quite a pity, sir, that Mr. Adams did not reserve his case and the remarks he had to make upon it until the Association meeting. He might then have made his remarks as to suppuration and no suppuration agree.

Mr. Adams performed his operation with a saw " three-eighths of an inch in width, with one inch and a half cutting edge, at the end of a small shank three inches in length."

I have before me, whilst I write, two specimens of bony anchylosis at the hip, from St. George's Hospital Museum. They are marked third series—III and IIIA, and they are the finest specimens I know. The following are the descriptions from the Museum catalogue.

" III. Portion of the right side of the pelvis, showing complete bony anchylosis of the hip joint in an extremely unnatural position, the femur being directed upwards. The bony union is quite complete, and there is also complete fusion of the anterior inferior spine of the ilium with the trochanter minor and neighbouring part of the femur, so that the femur is united to the whole front of the ilium, with the exception of a small oval aperture which still exists between the bones. The specimen was removed from the body of a young woman who died in the hospital of disease unconnected with the joint."

" IIIA. Innominate bone and part of femur, showing bony anchylosis in an unnatural position, the femur being directed forwards. The union is quite complete.   The preparation was removed from a man, aged 23, who had suffered from disease of the hip for fifteen years."

In the latter specimen, the circumference of the anchylosis is six and a half inches, and in the former it is six inches and one-eighth.   In both instances the new deposit is as hard as ivory.   The top of the trochanter occupies a position higher than that previously occupied by the head of the bone, and the neck of the bone is almost absorbed.   In the Hunterian Museum at the College of Surgeons there are four specimens, marked respectively 3325, 3326, 3327, and 3327 B, which differ only slightly, if at all, from those I have already described.

Having described the width of his saw, before the Association meeting, as three-eighths of an inch, Mr. Adams describes it in his letter to you (February 18) as " a quarter of an inch."   Now, one-eighth of an inch is important in such an operation.   We may suppose three-eighths of an inch to be the correct measure.   Of what size must the external opening be to allow a saw three-eighths of an inch in width so to work as to divide bone six inches in circumference ?   It may be that union has so taken place that a straight section may be made.   Sometimes, however, it is otherwise, as it was in the case on which I operated at Brighton, and as it is represented in the specimen III to which I have already alluded.   With such a disposition of the parts it is absolutely necessary to alter the direction of the saw when the bone is in part cut through ; but this can only be done by having an external opening equal to such alteration of direction.   This fact has been recognised by those who have practised and written on subcutaneous osteotomy long before Mr. Adams' attention was directed to the subject; and they have acknowledged the necessity of varying the size of the external opening according to the circumstances of the case.

But, on referring to the *Journal* of December 24th, 1870, we find Mr. Adams' view of the neck of the thigh-bone, as he is supposed to have cut through it.   Of course there can be no difficulty in cutting through the neck of the bone, unaltered as it is there described.   Whilst that shape was preserved, it is very unlikely that bony anchylosis had taken place.   But this operation was instituted for bony anchylosis.   From the description given by Mr. Adams, it is clear that very little alteration had taken place in the neck of the bone.

I examined this patient, Luke Bristowe, at the Royal Orthopædic Hospital, when he was under the influence of chloroform, and I came to the conclusion, and expressed my opinion that it was a case of rigid fibrous anchylosis. I confess that I was astonished to hear that in such a case section of the femur was resorted to.

I am happy, Sir, to think that I did not commence this correspondence. I should not have taken any notice of the case if Mr. Adams had not attacked me; but I think it will now be seen that his "perfectly new and original operation" for subcutaneous osteotomy might perhaps with advantage have been dispensed with.

I regret that it has been necessary to write such a long letter; but, before I conclude, I must ask you to correct a slight error in my last letter. In a quotation from my case of subcutaneous division of the neck of the thigh-bone, the following occurs : " The wound healed in *about* its entire extent." It should be, " The wound healed in almost its entire extent by the first intention, and in three weeks it was firmly cicatrised, so that passive motion could be freely employed."

I am, &c.,
B. E. BRODHURST,

London, March 7, 1871.

NO. 5.   LETTER FROM MR. ADAMS.

(Reprinted from the " British Medical Journal," March 25th, 1871).

SIR,—A few lines will suffice to answer Mr. Brodhurst's lengthy letter. It is clear, now, that Mr. Brodhurst never has performed subcutaneous section of the neck of the femur, an operation which I performed, for the first time in the history of surgery, on the 1st of December, 1869, with instruments specially devised for the purpose, and which I described and figured in your Journal of December 24th, 1870. This operation has since been successfully performed by the method, and with the instruments which I devised, by Mr. T. R. Jessop of Leeds in two cases, by Mr. Furneaux Jordan of Birmingham, and by Mr. Jowers of Brighton.

In Mr. Brodhurst's published lectures in 1869, he omitted all reference to any operation of this kind. In republishing his lectures in his book on Deformities, subsequently to the reading of my paper at the British Medical Association at Newcastle in August, 1870, he inserted a passage, p. 152, in which he appeared to claim priority of perfor-

F

mance of my operation. It seems, however, that the operation mentioned by him was on the great toe, and Mr. Brodhurst has now
abjured any claim of priority based upon it.

His claim, then, is now reduced to that which can be based upon
an operation which he performed on the hip joint in the year 1861,
not at all of a subcutaneous character, or so described by himself in
his various references to it; either in the *Proceedings* of the Royal
Medical and Chirurgical Society, vol. iv, p. 97, or in his work on
Deformities, p. 150, where it is also described, and placed in another
category altogether, side by side with one of Dr. Barton's for establishing a false joint. This operation he performed " by making,
in the first instance, an external incision to the extent of two inches
and a half," then extended to five inches, so as to reach from the
great trochanter to Poupart's ligament; through which extensive
opening Mr. Brodhurst states, in his book on Deformities, p. 151,
" I cut through the neck of the femur immediately below the head of
the bone, and then gouged away the remains of the head and the
dead bone from the acetabulum." To call this operation *subcutaneous*,
requires no small courage. It is, as I have said, an afterthought
subsequent to the publication of my case, and even of more recent
date than Mr. Brodhurst's lately published book; and I must be
pardoned for saying it does more credit to his ingenuity than to his
candour.

I am, etc.,

WILLIAM ADAMS.

Henrietta Street, Cavendish Square, March 20th, 1871.

NO. 6.   LETTER FROM MR. BRODHURST, WITH REMARKS BY
THE EDITOR CLOSING THE CORRESPONDENCE.

Reprinted from the British Medical Journal, April 1st, 1871.

SIR,—I did not press my argument in my last letter to its legitimate
conclusions; first, because I hoped it might not be necessary to do so ;
and secondly, because, if necessary, I desired to illustrate my meaning
with the aid of diagrams.

Fig. 1 is taken from the specimen III, to which I referred in my last letter. (Here follows a drawing of the specimen III from St. George's Hospital above referred to, and in which the operation could not have been performed in consequence of destruction of the head and neck of the bone, and broad surface of anchylosis, also from the quantity of new bone thrown out. This specimen very much resembles the one in St. Thomas' Hospital, No. 51'D, represented in woodcut Pl. III. Fig. C. W.A.) It shows a portion of the right side of the pelvis, as seen from behind, the pelvis being held in its normal position, with bony anchylosis of the hip joint, and the femur directed upwards and forwards. This specimen represents a somewhat similar deformity to that on which I operated at Brighton in 1861, and the line *a c* shows the line of section of the neck of the thigh-bone as it was then performed. Thus it will be seen that at a certain point the direction of the saw was changed, and was made to follow the line *a b*, for the bone could not otherwise have been cut through, any more than the specimen to which I now refer could otherwise be cut through. But it is obvious that the external incision must be sufficiently long between *b c* to allow this change in the direction of the saw to be made. In my case it required that this incision should be two and a half inches in length. It is equally obvious that if the external incision be made between *b c*, the section *a c* must be subcutaneous. This was the operation which I planned, and of which I gave the description before the Royal Medical and Chirurgical Society.

Fig. 2 is an exact copy of diagram No. 2, in Mr. Adams' communication to the Association in the *Journal* of December 24th, 1870. The

FIG. 10.

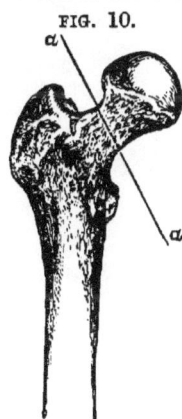

following description accompanies it : " Upper portion of thigh-bone. Situation and direction of subcutaneous division of neck of thigh-bone, represented by line *a a*."

Thus it will be seen that Mr. Adams made his section, as shown in the diagram, when little or no alteration had taken place in the neck of the thigh-bone. If bony anchylosis had taken place, the section could not have been made as it is described and represented by Mr. Adams, for the sufficient reason that the parts are not there. The conclusion, therefore, at which I am forced to arrive is, that my opinion as expressed at the examination of the patient is confirmed, and that Mr. Adams operated on a case of fibrous anchylosis by making a section of the neck of the bone.   But, it may well be asked, is such an operation justifiable ?

And now, Sir, I gladly stop.   It may be a satisfaction, however, to Mr. Adams to know that in the forthcoming number of the *St. George's Hospital Reports* I have contributed an article on anchylosis.   This article was written before my book (which has been the cause of so great offence to Mr. Adams) was published, and it was in Dr. Ogle's possession, and was, I believe printed before this correspondence commenced.   There I have described my operation on the hip in 1861, as it was described before the Royal Medical and Chirurgical Society, as subcutaneous.   I shall have much pleasure in sending Mr. Adams a copy of this paper, and he will then have an opportunity of making the history of subcutaneous osteotomy more complete, and of correcting some errors which have doubtless through ignorance or inadvertence crept into his writings on this subject.

I am, &c.,
B. E. BRODHURST.

March 27th.

This correspondence must end here.   In closing it, however, we think it right to express the opinion that Mr. Adams has added an original and valuable operation to the resources of surgery, and that the details which Mr. Brodhurst gives of his interesting case do not affect that claim unfavourably—ED. " B. M. J."

END.

www.ingramcontent.com/pod-product-compliance
Lightning Source LLC
Chambersburg PA
CBHW021951190326
41519CB00009B/1217